Lecture Notes in Management and Industrial Engineering

Series Editor

Adolfo López-Paredes, INSISOC, University of Valladolid, Valladolid, Spain

This book series provides a means for the dissemination of current theoretical and applied research in the areas of Industrial Engineering and Engineering Management. The latest methodological and computational advances that can be widely applied by both researchers and practitioners to solve new and classical problems in industries and organizations contribute to a growing source of publications written for and by our readership.

The aim of this book series is to facilitate the dissemination of current research in the following topics:

- Strategy and Entrepreneurship
- Operations Research, Modelling and Simulation
- Logistics, Production and Information Systems
- Quality Management
- Product Management
- Sustainability and Ecoefficiency
- Industrial Marketing and Consumer Behavior
- Knowledge and Project Management
- Risk Management
- Service Systems
- Healthcare Management
- Human Factors and Ergonomics
- Emergencies and Disaster Management
- Education

More information about this series at https://link.springer.com/bookseries/11786

Rahul S Mor · Sachin S. Kamble ·
Kuldip Singh Sangwan
Editors

Operations and Supply Chain Management in the Food Industry

Farm to Fork

Springer

Editors
Rahul S Mor
National Institute of Food Technology
Entrepreneurship and Management
Kundli (Sonepat), India

Sachin S. Kamble
EDHEC Business School
Roubaix, France

Kuldip Singh Sangwan
Birla Institute of Technology and Science
Pilani, Rajasthan, India

ISSN 2198-0772 ISSN 2198-0780 (electronic)
Lecture Notes in Management and Industrial Engineering
ISBN 978-981-16-5554-8 ISBN 978-981-16-5555-5 (eBook)
https://doi.org/10.1007/978-981-16-5555-5

This Springer imprint is published by the registered company Springer Nature Singapore Pte Ltd.
The registered company address is: 152 Beach Road, #21-01/04 Gateway East, Singapore 189721, Singapore

Foreword

Food supply chains across the globe are operating under increasing pressure to keep feeding the world's growing population. In this context, one of the most important challenges faced by operations and supply chain professionals concerns how to improve the sustainability of food supply chains, while simultaneously managing the competitiveness of complex food systems. In the outstanding book *Operations and Supply Chain Management in Food Industry: Farm to Fork*, several of the most talented researchers in the fields of food operations and supply chains provide readers with unique insights into the ways in which emerging technologies, such as blockchain, can unlock more sustainable food supply chains in order to achieve the challenging target of feeding the globe while contributing to meeting Sustainable Development Goals. The book has been carefully edited by some of the most respected researchers in the field, with many years of both practical and academic experiences. It is a must-read book for any scholar, practitioner, or food policymaker who is committed to creating a fairer and more sustainable world.

Charbel Jose Chiappetta Jabbour
Global Chair Professor in Sustainable
Supply Chain and
Operations Management
Lincoln International
Business School (LIBS)
University of Lincoln
Lincoln, UK

Professor, Emlyon Business School
Ecully-Lyon, France

Preface

The Indian food supply chain is highly complex, involving government and private traders, big and small players, traditional as well as modern systems. All parts of the Indian food supply chain—procurement, storage, transportation, processing, packaging, and delivery—involve different systems and are not dependent upon a single system. It partially explains why the Indian food supply chain shows strong resiliency during the current pandemic lockdown. Food waste/spoilage and pilferage need to be brought down during post-harvesting, harvesting, and pre-harvesting to feed the growing world population at a reasonable cost. The agility and traceability in food supply chains are critical factors to be improved using modern techniques like blockchain and data analytics. However, more research is required to study these losses, their quantification, root causes, and mitigation techniques. The requirement of modern and digital technologies in storage, processing, and packaging; better coordination among government agencies, traders, and farmers; and improved IT and transportation infrastructures provide entrepreneurial opportunities for the IT savvy young Indians in this field. This book essentially presents these aspects of the food supply chain in the Indian context.

Chapter 1 highlights the essence of an agile agricultural supply chain management leading to higher productivity, profitability, and customer satisfaction. A perishable food supply chain is analyzed in Chap. 2 to identify the risk factors involved in the storage and transportation of perishable foods. In Chap. 3, forecasting methods are illustrated by a case study to provide an accurate judgment to the people involved in food supply chain management. The procurement and value chain models of the black pepper market are identified in Chap. 4, where identification and mapping of stages involved, intermediaries' functions, and value chain mapping thus lay the foundation for the study. In Chap. 5, a study has been conducted at a coffee and bakery shop in Bangkok, Thailand, to understand the supply chain and various food safety measures being taken at its scale and highlight the food supply chain threats. The need to identify, understand, and resolve the traceability and transportation issues for an efficient and responsive food supply chain is highlighted in Chap. 6. The critical success factors that can facilitate the successful adoption of blockchain technology in the food supply chain are presented in Chap. 7. The identified hierarchical structure and contextual relationships among these factors are expected to help policymakers drive the use of blockchain technology in this area. Chapter 8 highlights the sustainability enablers in the agricultural supply chains, and these sustainability enablers are mapped with the United Nation's Sustainable Development Goals (UNSDGs). The application of digital technologies such as cloud computing for better and effective services in a food supply chain is discussed in Chap. 9. Chapter 10 describes various green technologies for better safety and quality in the food supply chains. Chapter 11 entails that a communication-based packaging system is an important technique that improves or indicates product safety in the supply chain using smart devices in the food packaging system. Finally, Chap. 12 provides an entrepreneurial policy framework and guidelines for developing agri-food entrepreneurship with successful case studies.

Sonepat, India Rahul S Mor
Roubaix, France Sachin S. Kamble
Pilani, India Kuldip Singh Sangwan

Acknowledgements

We acknowledge all those people who were involved and helped in completing this book project. Firstly, we would like to thank the authors for contributing their valued time and expertise. Special thanks are due to the reviewers' valuable contributions regarding the improvement of quality, coherence, and content demonstration of the chapters. We also appreciate the referees for reviewing the manuscripts and scholars for editing and organizing the chapters. Finally, the editors are grateful to their parent institutes (National Institute of Food Technology Entrepreneurship and Management, Kundli (Sonipat) - 131028, India; EDHEC Business School, Roubaix, France; Birla Institute of Technology and Science, Pilani - 333031, India) for providing essential facilities to conduct this research work smoothly.

Rahul S Mor
Sachin S. Kamble
Kuldip Singh Sangwan

Contents

Editors and Contributors

About the Editors

Dr. Rahul S Mor is a faculty and researcher of operations management, supply chain & logistics, industrial engineering & management, manufacturing systems, etc., and has over four years of experience, primarily at NIFTEM Kundli. He holds a Ph.D. in Industrial and Production Engineering, specializing in Supply Chain & Operations Analytics. He has been teaching and mentoring the students of UG & PG courses, and has also taken the lead in various academic, outreach and administrative assignments. Dr. Mor has authored/co-authored over 50 publications in international refereed SCI, ABDC, CABS ranked journals of high impact factor, & many chapters, books, conference papers. One of his MoFPI (Govt. of India) funded research projects supports the agri-food supply chain digitalization through disruptive technologies, among other R&C projects in the manufacturing and agri-food sector. He is supervising research at the doctoral and masters levels. Dr. Mor has edited various books on 'Agri-Food 4.0, 'Circular Economy', 'OSCM', 'Industry 4.0' with Springer, Emerald, EAI-Springer, etc. He is the Area Editor and Managing Guest Editor: Opr. Mgmt. Res. (Springer), Int. J. Logi. Res. & Appl. (T&F); Guest Editor: IEEE Trans. on Engg. Mgmt. (IEEE), Int. J. Transp. Eco.; Associate Editor: Supply Chain Forum: An Int. J. (T&F); Editor: IJSOM; Editorial Board: JDS, FSFS, IJCEWM, IJLT, etc., and reviewer for numerous int. journals. He is a member of many professional societies.

Dr. Sachin S. Kamble is Professor of Digital Transformation at EDHEC Business School, France. He holds a Ph.D. in Management, MBA in Operations, and bachelor's degree in Mechanical Engineering. In academics, he has been associated with various institute of repute such as National Institute of Industrial Engineering (NITIE) Mumbai, Hyderabad Central University and Karnatak Univeristy. Dr. Sachin's teaching and research interests include Operations Management, Supply Chain Management, Big Data Analytics, Industry 4.0, and Blockchain Technology.

He has more than 50 authored/co-authored publications in referred international journals and includes high-impact journals such as *International Journal of Production Economics, International Journal of Production Research, Computers in Industry, Computers and Operations Research, International Journal of Information Management* etc. Dr. Sachin has completed 27 consultancy assignments for leading manufacturing organizations that include TNT Logistics Ltd., Parle Limited, ACC Ltd., Intas Pharma, Mundra Ports and SEZ Ltd., Indogulf Fertilisers Ltd., TTK-LIG Ltd, BPCL-Mumbai, Gujarat State Fertiliser Corporation, Kochi Refinery, Indian Rayon, RCF Ltd, AICTE and Glaxo Smithkline. He has also conducted various MDP's for the senior-level executives from ACC Ltd., HDFC Bank, BEL, HMT, L&T, Apollo Tyres Ltd, TMT Ltd., etc. in Operations and SCM. Dr. Sachin has also conceptualized special certificate program on SC and Logistics Management for Larsen and Toubro (one-year duration) and Manufacturing Excellence Program for Apollo Tyres Limited.

Dr. Sachin is presently guest editing special issues for *International Journal of Production Research* in Industry 4.0 and supply chain finance, and for *Journal of Global Operations and Strategic Sourcing* on the applications of Blockchain technology in global operations.

Dr. Kuldip Singh Sangwan is a Senior Professor of Mechanical Engineering at Birla Institute of Technology and Science Pilani (BITS Pilani), Pilani campus. He is the recipient of prestigious Shri B. K. Birla and Shrimati Sarala Birla Chair Professorship. He is an active researcher in the field of sustainable manufacturing, green supply chain management, energy-efficient machining, industry 4.0, life cycle engineering, and application of artificial intelligence in manufacturing system design. He has to his credit 13 completed, 8 ongoing PhDs, three edited books and more than 160 research papers in journals like IJPR, PPC, IJOPM, JCLEPRO, Ann. Oper. Res., IJLCA, IJAMT, JMTM, etc. He has established Joint Indo- German Experience Lab, Sustainable Manufacturing & Life Cycle Engineering Lab, and Advanced Precision Lab at BITS Pilani, Pilani Campus. He has projects worth more than 0.9 million € with international agencies (DAAD, IGES), industry and Indian Government agencies. He is a fellow of the Institution of Engineers (India), and a life member of Indian Institute of Industrial Engineering (IIIE) and Society of Operations Management (SOM). Professor Sangwan is the current editor of Industrial Engineering Journal published by Indian Institute of Industrial Engineering. His named has featured in the top 2% world scientists in the area of Industrial Engineering and Automation published by Stanford University researchers.

Contributors

Rajeev Agrawal Department of Mechanical Engineering, Malaviya National Institute of Technology, Jaipur, India

Sapna Arora Department of Food Business Management and Entrepreneurship Development (FBMED), National Institute of Food Technology Entrepreneurship and Management (NIFTEM), Kundli, Sonipat, Haryana, India

Manisha Bansal Indian Institute of Technology Delhi, New Delhi, India

Rohitash Kumar Banyal Department of Computer Sciene and Technology, Rajasthan Technical University, Kota, India

Ashish Dwivedi Jindal Global Business School, O. P. Jindal Global University, Sonipat, India

Prayag Gokhale Department of MBA, KLE Dr. M S Sheshgiri College of Engineering and Technology, Belagavi, Karnataka, India

Disha Gulia Department of Food Business Management and Entrepreneurship Development (FBMED), National Institute of Food Technology Entrepreneurship and Management (NIFTEM), Kundli, Sonipat, Haryana, India

Gajanand Gupta School of Mechanical Engineering, Vellore Institute of Technology, Chennai, Tamil Nadu, India

Darshana Hooda University Computer Center, Deenbandhu Chhotu Ram University of Science and Technology, Murthal, India

Thappasarn Jaikaew School of Environment, Resources and Development, Asian Institute of Technology, Bangkok, Thailand;
Department of Food Technology and Nutrition, Faculty of Home Economics Technology, Rajamangala University of Technology Krungthep, Bangkok, Thailand

Sarat Kumar Jena Xavier Institute of Management, Bhubaneswar, India

Ajay Jha Operations Management, Jaipuria Institute of Management, Lucknow, India

Anil Jindal Department of Mechanical Engineering, Giani Zail Singh Campus College of Engineering and Technology, Bathinda, Punjab, India

Sachin S. Kamble EDHEC Business School, Roubaix, France

Anand Kishore Department of Food Engineering, National Institute of Food Technology Entrepreneurship and Management, Kundli (Sonepat), India

P. M. Kulkarni Department of MBA, KLS Gogte Institute of Technology, Belagavi, Karnataka, India

Pushpendra Kumar Department of Food Engineering, National Institute of Food Technology Entrepreneurship and Management, Kundli (Sonepat), India

Vijay Kumar Department of Basic and Applied Science, National Institute of Food Technology Entrepreneurship and Management, Kundli (Sonepat), India

Khushbu Kumari National Dairy Research Institute, Karnal, India

Modisa School of Environment, Resources and Development, Asian Institute of Technology, Bangkok, Thailand

Rahul S Mor Department of Food Engineering, National Institute of Food Technology Entrepreneurship and Management, Kundli (Sonepat), India

Tribhuvan Nath Institute of Management Studies, RGSC, Banaras Hindu University, Varanasi, Uttar Pradesh, India

Neeraj Pandey National Institute of Industrial Engineering (NITIE), Mumbai, India

Anupama Panghal Department of Food Business Management and Entrepreneurship Development, National Institute of Food Technology Entrepreneurship and Management (NIFTEM), Sonipat, Haryana, India

Akshay Patidar Department of Management Studies, Malaviya National Institute of Technology, Jaipur, India

Kumar Rahul Department of Basic and Applied Science, National Institute of Food Technology Entrepreneurship and Management, Kundli (Sonepat), India

Kuldip Singh Sangwan Department of Mechanical Engineering, Birla Institute of Technology and Science Pilani, Pilani, India

Avanish Shandilya Department of Food Business Management and Entrepreneurship Development (FBMED), National Institute of Food Technology Entrepreneurship and Management (NIFTEM), Kundli, Sonipat, Haryana, India

Monica Sharma Department of Management Studies, Malaviya National Institute of Technology, Jaipur, India;
Department of Mechanical Engineering, Malaviya National Institute of Technology, Jaipur, India

Rohit Sharma Jaipuria Institute of Management, Noida, India

Satyendra Kumar Sharma Department of Management, Birla Institute of Technology and Science Pilani, Pilani, India

Anjali Shishodia LMTSM, Thapar University, Patiala, India

Anupama Singh Department of Food Engineering, National Institute of Food Technology Entrepreneurship and Management, Kundli (Sonepat), India

Harsh Srivastava Organization Design, GMR Group, New Delhi, India

Priyanka Vern Department of Food Business Management and Entrepreneurship Development, National Institute of Food Technology Entrepreneurship and Management (NIFTEM), Sonipat, Haryana, India

Vikas Yadav Indian Railway, New Delhi, India

Owais Yousuf Department of Bioengineering, Integral University, Lucknow, India

Chapter 1
Agile Agriculture Supply Chain Management (AASCM) for Managing Shifting Consumer Food Preferences: Framework Development Using Grounded Theory Approach

Neeraj Pandey

Abstract Consumer food preferences are changing. These changes have both temporal and spatial dimensions. An agile agriculture supply chain management (AASCM) is required to cope up with this shifting trend in food consumption. The AASCM would ensure an uninterrupted supply of agricultural produce from farm to fork. This would also contribute to the Sustainable Development Goal (SDG) of achieving food security, leading to zero hunger. This study adopts an inductive grounded theory approach to develop an AASCM framework to match the changing consumer food preferences seamlessly. This is the first attempt to develop the AASCM framework in the supply chain literature. The results showed that consumer insights, market sensing, decisiveness, state-of-the-art IT infrastructure, top management support, and strong vendor relationship are essential for achieving AASCM. The implementation of the framework will aid agility in supply chain management produce leading to higher agriculture productivity, profitability, and customer satisfaction. The study is useful for supply chain management professionals, agriculture entrepreneurs, and policymakers in the agri-food domain.

Keywords Agile agriculture supply chain management (AASCM) · Food preferences · Agriculture · Farm to fork · Agile supply chains · Grounded theory

1.1 Introduction

The twenty-first century has witnessed a growing interest in "agility." This is evident from the number of articles and research papers published on this topic. According to Google Scholar search results, more than 0.22 million research papers have been published on agility and related topics since 2000. The popularity of research on

N. Pandey (✉)
National Institute of Industrial Engineering (NITIE), Vihar Lake Road, Powai, Mumbai 400087, India
e-mail: npandey@nitie.ac.in

R. S Mor et al. (eds.), *Operations and Supply Chain Management in the Food Industry*, Lecture Notes in Management and Industrial Engineering,
https://doi.org/10.1007/978-981-16-5555-5_1

"agility" may be estimated from the fact that it took just twenty years to publish 45% of articles on agility, whereas the other 55% came in more than a hundred years. Agility has taken center stage due to the growing unpredictability in the business environment. The volatility was brought by new competitors in the industry, change in regulations, smaller product life cycles, and unpredictable demand and supply situations. The recent COVID-19 pandemic underlined the need for agile supply chain management for products and services. Organizations like Asian Paints, Mahindra Tractors, CEAT Tyres, and Uber Eats, which were more agile to adapt to the changing situations, were able to survive and grow during the COVID-19 pandemic [1].

There is a need for agile agriculture supply chain management (AASCM) given the industry's dynamic nature, including changes in the ongoing consumer food preferences. The agility would build flexibility in AASCM to respond to the customer and markets' requirements in a seamless manner. The agriculture supply chain management has also embraced technologies like blockchain [2, 3], machine learning [4], and the Internet of things (IoT) [5, 6]. The government of many countries has brought in reforms to modernize the agriculture supply chain management and free it from shackles of archaic rules. For example, the Government of India has passed Farms Bill in 2020, which would provide freedom to farmers for choosing any kind of agriculture supply chain management, pricing, and storage facilities.

The changes in consumer food preferences had a direct impact on agriculture supply chain management. The shift in consumption basket has altered the agriculture supply chain management, including the burden on the cold supply chain facilities. Consumers have become more aware of their health [7–9]. This awareness has an impact on food practices. The changing food preferences may be due to lifestyle changes, food trends, cultural or psychological factors [10]. There is a growing intention to purchase organic food products among the consumers in the emerging markets [11]. The safety concerns regarding quality are also making consumers alter their food habits [12]. These shifts in food consumption require AASCM to provide uninterrupted supplies from farm to fork. There are hardly any studies that have captured agility issues in the agriculture supply chain domain. The remaining paper is organized as follows. The next section has a literature review followed by the methodology, data collection, analysis, and results. The final section consists of discussions and conclusions besides highlighting the limitations and future scope of work on AASCM.

1.2 Literature Review

Consumers have got multiple food options. This information about packaged food ingredients, advertisements about organic cereals, and awareness about healthy food habits is making people shift their food preferences. Many organizations get patent on novel food items to safeguard against imitation of the same product by other competing firms [13]. A large study in Poland found that people were shifting to novel healthy foods. This has led to improve state health indicators. There was a growing interest of citizens for advertisements having positive health content [14]. The customers especially Millennials have become more vocal about their various likings including food preferences on social media sites like Facebook [15, 16]. The consumers in emerging markets like India were preferring fresh food items over packaged food. The purchase from nearby small stores was preferred to buy fresh food [17].

The changing food preferences of the population were evident from consistently declining lamb consumption in Canada and USA. Stanford et al. [18] found that this change in food consumption habits was due to the rising supply chain cost, availability of cheaper food alternatives, and lack of government support to this sector. The supply chain has to cater to the retail outlets, which have witnessed a shift in consumer demand for products [19]. There has been a growing interest of consumers in green commodities and organic food products [20, 21]. This has led to forced adjustments in local supply chains. The products in high demand and those with higher margins, especially bought by wealthier consumers, were given high priority. These products were given precedence in supply chain planning and execution [19].

Aging is another factor contributing to change in food preferences [22, 23]. This has been witnessed in both the younger and older population. The European Union's aging population has become more conscious of food's health and nutritional values [23]. The growing preference for local and organic food in the USA also changed the supply chain networks in North America [24]. This willingness to pay (WTP) is an important parameter to measure actual purchase behavior [25, 26]. Adams and Salois [24] found that there was a high WTP by USA consumers for local foods. This highlighted that the supply chain needs to realign to adjust to the paradigm shift in food preferences.

Zu et al. [27] proposed operation research (OR) models as one of the solutions to realign the supply chains to manage these changes in food preferences. However, few of the supply chain issues cannot be empirically modeled as the real world is too complex and may require a hybrid approach. Various agri-food organizations, including US poultry industry, had adopted vertical coordination in supply chain management by leveraging contracting and technology [28]. Various authors [29, 30] have emphasized that AASCM should also ensure that there should be traceability in the agriculture supply chains.

1.3 Methodology

The study uses inductive grounded theory approach to develop a framework on agile agriculture supply chain management (AASCM) in the context of changing food preferences across the globe. The grounded theory is useful in developing frameworks where there is a lack of sufficient literature. It helps build a fresh perspective and bring in new technologies, context, and constraints that were not considered in the earlier models [31, 32]. The grounded theory helps construct a framework from the first data set and iterates it with the subsequent data [33]. The exercise continues till the data is saturated, and there is no new insight from the fresh responses. The grounded theory has its origin in pragmatism and symbolic interactionism [33]. This study collected data from supply chain professionals. The respondents had experience in managing supply chains in the agri-food industry, tire, fast-moving consumer goods (FMCG), chemicals, and pharmaceutical industry. The industry diversity of respondents brought fresh perspectives in developing a framework for agile agriculture supply chain management (AASCM).

1.4 Data Collection and Analysis

The first respondent had more than seven years of experience in agri-food supply chain management in an MNC. The respondents were not revealed the actual purpose of the exercise to avoid any social desirability bias. The questions started with the state of the industry and issues with the supply chain and led to discussions regarding changing consumer preferences, agility in supply chains, coping mechanism, building agile agriculture supply chain management, etc. The literature review helped build the initial set of questions for grounded theory research. The subsequent interview questions used the knowledge gained in the previous interactions regarding issues in agriculture supply chain management. The theoretical memos were jotted in each session with the respondents. The analysis started just after collecting data from the first respondent. The snowballing technique was used to get a reference from the first respondent regarding other supply chain professionals. The responses were analyzed using content analysis. Each respondent had a unique identity number. The analysis involved open coding, axial coding, and selective coding, as suggested by Corbin and Strauss [33]. The saturation of responses was achieved after twenty-three responses.

1.5 Results

The key factors for achieving AASCM emerged from the grounded theory exercise. Table 1.1 provides the six AASCM key dimensions and three outcome variables. The outcome variables are goals that the agri-food company chase while building agile supply chains. The six input factors for AASCM are:

Consumer Insights: The consumer insights not only help to know about the current consumer preferences but also provide knowledge about how food preferences are changing. It tracks consumer food purchase baskets, food search habits, and willingness to pay for different food items. It relies heavily on data to provide consumer insights so that the supply chains may realign their resources to meet the changing

Table 1.1 AASCM key dimensions

S. No.	Sample memos [respondent ID]	Open coding	Axial coding	Selective coding
1	We keep a close eye on the market through various databases and use our marketing intelligence inputs. It gives us a sense of demand patterns, and accordingly, we align our supply chain to meet the predicted demand [7]	Data (3)	Data	Consumer insights
		Information (3)		
		Software (2)		
		Insights (3)	Insights	
		Consumer behavior (3)		
		Intelligence (1)		
		Predictions (3)		
2	My company has one of the best IT systems and software. It helps us to leverage various data opportunities to predict demand shifts. I believe that the latest digital technologies should be used by supply chain professionals. Soon, drone and sensors would become common in agri-food supply chain management [15]	Business (9)	Marketplace	Market sensing
		Market (1, 3, 10)		
		Demand (1,2, 4)		
		Intelligence (1)	Visualize change	
		Demand patterns (1)		
3	We have a heavy focus on data and use it for market predictions. The changing consumer food consumption behavior is impacting the supply chains. The food industry is now using data from various sources to get consumer insights. We also receive useful information from our dealers and other channel partners [23]	Choices (4)	Decisions	Decisiveness
		Decision (4)		

(continued)

Table 1.1 (continued)

S. No.	Sample memos [respondent ID]	Open coding	Axial coding	Selective coding
		Changes (4)	Agile	
		Responsive (4)		
		Timely (6,7, 10)		
4	There are many food choices for consumers. The consumer food preferences are changing. The supply chain should be agile to be responsive to these changing demands. The decision making should be data-driven and fast to match the demand changes [12]	IT (2)	IT infrastructure	State-of-the-art IT infrastructure
		Digital (2)		
		Drone (2)	Technology	
		Latest technology (2)		
		Sensors (2)		
5	Today, the competition is not between company A and company B but between its value chains. A close rapport with the vendor besides their training is a must in this competitive world. The close cooperation with vendors helps us in getting through the thick and thin of the business [9]	Top management (7)	Management support	Top management support
		Support (7)		
		Ownership (7)		
		Resource availability (7)	Facilitation	
		Funding (7)		
6	The food category B2B customer looks for quality, variety, and traceability through existing supply chains. If the customers get the desired quantity and quality on time, the buyers are satisfied [14]	Close rapport (5)	Relationship	Strong vendor relationship
		Training (9)	Vendor development	
		Cooperation (5)		
7	Our agility in the supply chain management is due to the top management support. We are able to deliver in short lead times with varying volumes due to complete support and resource availability, including required funding by the management. The team feels ownership for their respective projects [4]	Quality (6)	Customer expectation	Customer satisfaction (B2B)
		Variety (6)		
		Traceability (6)		
		Quantity (6)	Delivery	
		Quality (6)		

(continued)

Table 1.1 (continued)

S. No.	Sample memos [respondent ID]	Open coding	Axial coding	Selective coding
		On time (6)		
8	Profit is the acid test for a company. Profit means revenue received from the customer minus the cost to serve that customer. In my view, agility in supply chains helps in reducing cost and enhancing revenue [19]	Year-on-year (10)	Growth	Organizational productivity
		Growth (10)		
		Expand (10)		
		Productivity (10)	Efficiency	
		Faster (10)		
		Delivery (10)		
10	The agile supply chains in agriculture would lead to higher productivity for all stakeholders. It helps to expand to the newer markets. It would mean faster on-time delivery with year-on-year growth [21]	Profit (8)	Profit	Company profitability
		Revenue (9)		
		Cost (8)	Cost to serve	
		Cost to serve (8)		

local and global demand. The respondents emphasized this as one of the critical inputs for building AASCM.

Market Sensing: The majority of the respondents emphasized that market sensing was very important in the quest for building agile agriculture supply chains. It is critical for agri-food supply chains as the food consumption patterns keep shifting. The shift has both temporal and spatial dimensions. The channel partner inputs have been helpful in connecting the dots about the demand shifts in agricultural products. The market sensing provides time to agriculture produce-based organizations to plan and reorient resources as per the consumption shifts in the marketplace. The market sensing also involves innovative ways of generating market intelligence like inputs from agriculture salesperson conferences, competitor social media campaign analysis, competitor new product development data analysis, channel partner inputs, international agriculture trade reports, competitor supply chain shifts, etc.

Decisiveness: The speed of decision making is a key aspect in developing AASCM. The support from the team members and top management is essential in taking and executing AASCM-related decisions. The decisiveness is a critical element of AASCM as agility would require bold and quick decisions as per the demand and supply of agricultural produce in a particular region.

State-of-the-art IT Infrastructure: The technologies of today have made it possible to plan and execute AASCM. Each respondent referred to technology, including techniques like machine learning, blockchain, optimization, and automation, in her/his discussions for bringing agility in given supply chains. The state-of-the-art IT infrastructure is an essential prerequisite for building a robust AASCM by leveraging technology.

Top Management Support: The support by top management is an essential condition for developing AASCM. The agile supply chains require a strong central team to plan and execute various demand scenarios. The top management support will ensure that the teams would get the required resources within the specified time besides flexibility to build and sustain agile agriculture supply chains.

Strong Vendor Relationship: The agri-food organizations not only are strengthening themselves but also focus a lot on developing strong vendor relationships to build agile agriculture supply chains. The vendors are the most credible source of market intelligence data that goes into building AASCM. The vendors are in touch with the farmers, consumers, competitors, and other large agri-food organizations on a day-to-day basis and have the latest information. The vendors ensure supplies, replacements, and update about market data. It helps in building agile agriculture supply chains. Therefore, the organizations invest in vendor development in terms of training, sharing of information, co-creation of products and services, and lending of resources on mutually agreed terms.

The outcome variables of AASCM are:

Customer Satisfaction (B2B): The AASCM aids in building a strong relationship with the B2B customers. It was witnessed that the organization which had an agile supply chain shifted their supply of agriculture produce raw material through trains when the road transport was stopped due to the pandemic. The liaison with local authorities helped in the transportation of these raw materials within the city. The agile supply chains minimized pandemic impact as the customer was able to receive raw material even in the time of crisis. This not only helped in achieving customer satisfaction but also helped in building a stronger business relationship with the buyers.

Organizational Productivity: The agility as a culture in the organization aids in enhancing the productivity of employees. The agile agriculture supply chains respond quickly to changing food consumption demands, thereby achieving organization targets and profitability. It also leads to catering to newer markets where there is a better opportunity to upsell the agri-food products. The implementation of blockchain and IoT is further adding to the agility and organizational productivity.

Company Profitability: The AASCM leads to higher profitability of an organization. The agility in agriculture supply chains leads to higher customer satisfaction. This leads to increased business and network for the organization.

1.6 Discussion and Conclusions

The study has developed a framework for building agile agriculture supply chain management (AASCM) for managing agri-food demand shifts. The agile supply chains would require relevant and timely information to reorient the planning and execution. This AASCM model would be sustainable only if there is a high degree of automation in demand planning and supply chain management. This agile digital supply chain would be highly data-driven decisions without any bias. The COVID-19 pandemic has reinforced the importance of building agility in the agriculture supply chain to ensure food security for the people.

The AASCM framework emphasizes on leveraging the power of data. This agri-food data would be drawn from various sources, including inputs from channel partners, trade reports, government agriculture and food-related reports, survey, company historical data, marketing intelligence, and experiment data. The data analytics output would help to get the customer insights about the upcoming agri-food demand shift. These insights go as input for resource planning and allocation for supply chain realignment for the coming week, month, or quarter (Fig. 1.1).

The AASCM would require strong top management commitment and close vendor involvement. The top management support would ensure timely allocation of resources, including requisite manpower, to the team involved in managing AASCM. The vendor support is essential in AASCM to ensure the timely availability of agri-food raw material, semifinished goods, and delivery of finished food products. Therefore, the development of close vendor relationship is crucial for building AASCM. The study also highlighted that the AASCM would help attain higher agriculture productivity, profitability, and customer satisfaction.

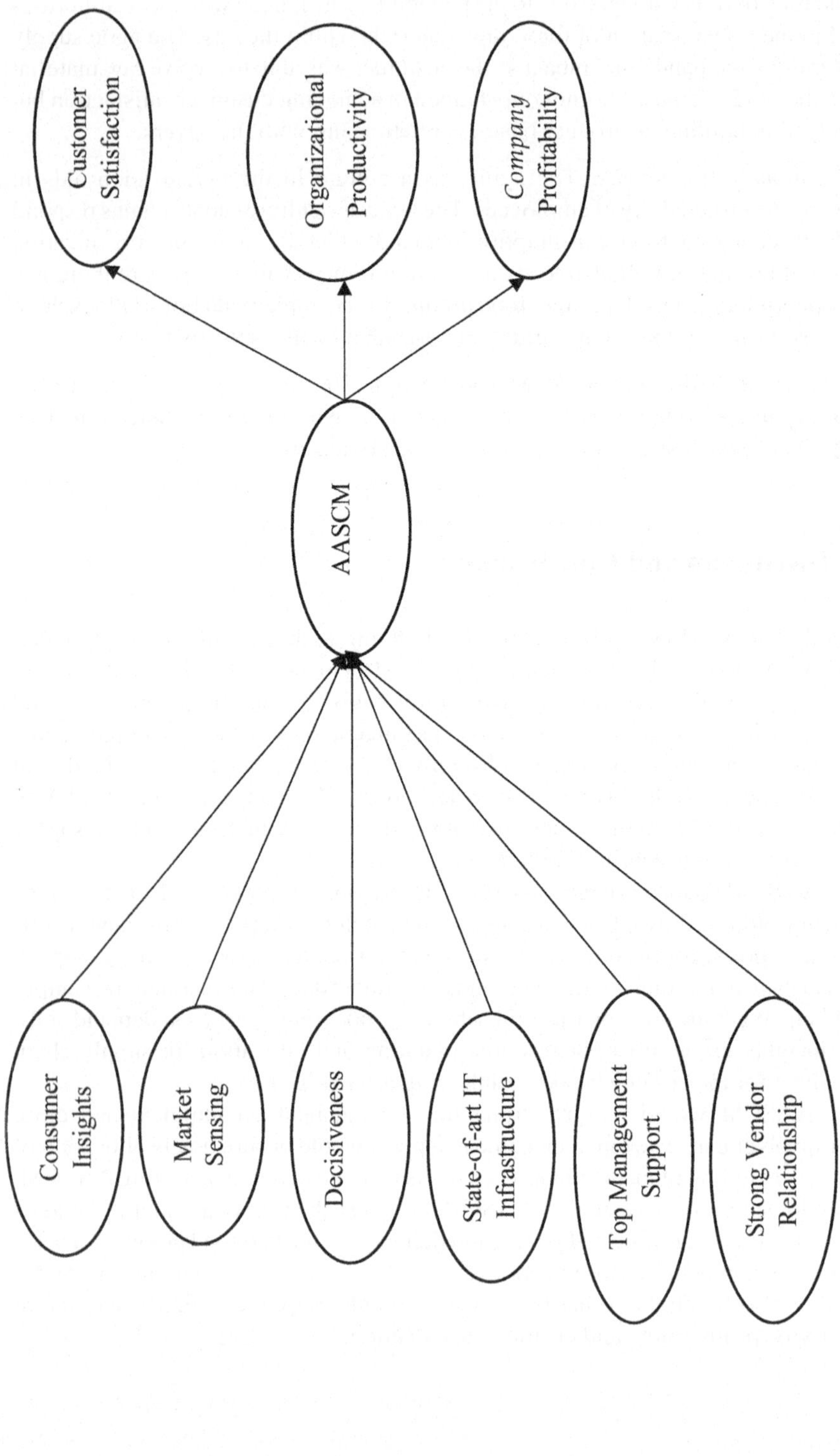

Fig. 1.1 Agile agriculture supply chain management (AASCM) framework for managing consumer food preferences *Source* Author

1.7 Limitations and Future Scope

This study has a few limitations, which may become an exciting work scope for future scholars. The study has focused on agriculture supply chains. Future researchers may adapt this model for developing agile supply chain management in other industries like chemical, pharmaceuticals, health care, e-commerce, and FMCG sector using grounded theory. The AASCM should be tested empirically using standard scales and primary data to refine it further. The role of the latest techniques like machine learning, blockchain technologies, IoT, cloud computing, and artificial intelligence (AI) in building AASCM should be explored by future researchers.

References

1. Pandey N (2021) Digital marketing strategies for firms in post covid-19 era: insights and future directions. In: Chaturvedi H, Dey AK (eds) The new normal: challenges of managing business, social and ecological systems in the post COVID-19 era. Bloomsbury Publishing, New Delhi, pp 107–124
2. Kamble SS, Gunasekaran A, Sharma R (2020) Modeling the blockchain enabled traceability in agriculture supply chain. Int J Inf Manage 52. https://doi.org/10.1016/j.ijinfomgt.2019.05.023
3. Kamble S, Gunasekaran A, Arha H (2019) Understanding the blockchain technology adoption in supply chains-Indian context. Int J Prod Res 57(7):2009–2033
4. Kamble SS, Gunasekaran A, Kumar V, Belhadi A, Foropon C (2020) A machine learning based approach for predicting blockchain adoption in supply Chain. Technol Forecast Soc Change 120465
5. Kamble SS, Gunasekaran A, Parekh H, Joshi S (2019) Modeling the internet of things adoption barriers in food retail supply chains. J Retail Consum Serv 48:154–168
6. Luthra S, Mangla SK, Garg D, Kumar A (2018) Internet of things (IoT) in agriculture supply chain management: a developing country perspective. In: Emerging markets from a multidisciplinary perspective. Springer, Cham, pp 209–220
7. Nayal P, Pandey N (2021) Role of social media in hospital branding: insights for marketing practitioners. In: New techniques for brand management in the healthcare sector. IGI Global, pp 1–16
8. Pandey N, Raina A (2019) Five decades of research in healthcare pricing: future directions for academia and policymakers. Int J Manag Pract 12(3):301–320
9. Pandey N, Thombal AE (2018) Why stents are expensive: drivers of high prices in emerging markets. Health Mark Q 35(4):298–312
10. Asp EH (1999) Factors affecting food decisions made by individual consumers. Food Policy 24(2–3):287–294
11. Pandey N, Kaushik S (2012) Factors affecting consumer's green product purchase decisions: an empirical approach. Int J Bus Compet Growth 2(4):341–356
12. Rimal A, Fletcher SM, McWatters KH, Misra SK, Deodhar S (2001) Perception of food safety and changes in food consumption habits: a consumer analysis. Int J Consum Stud 25(1):43–52
13. Pandey N, Dharni K (2014) Intellectual property rights. PHI Learning Pvt. Ltd.
14. Wadolowska L, Babicz-Zielinska E, Czarnocinska J (2008) Food choice models and their relation with food preferences and eating frequency in the Polish population: POFPRES study. Food Policy 33(2):122–134
15. Pandey N, Gudipudi B (2019) Understanding 'what is privacy' for millennials on Facebook in India. J Data Protect Priv 2(3):224–233

16. Pandey N, Jha S, Singh G (2020) Promotion of green products on Facebook: insights from millennials. Int J Manag Pract 13(3):275–294
17. Ali J, Kapoor S, Moorthy J (2010) Buying behaviour of consumers for food products in an emerging economy. Brit Food J 112(2):109–124
18. Stanford K, Hobbs JE, Gilbert M, Jones SDM, Price MA, Klein KK, Kerr WA (1999) Lamb-buying preferences of Canadian abattoirs and producer marketing groups: implications for the Canadian supply chain. Supply Chain Manage: An Int J 4(2):86–94
19. Codron JM, Grunert K, Giraud-Heraud E, Soler LG, Regmi A (2005) Retail sector responses to changing consumer preferences. New Direct Global Food Markets 794:32–46
20. Tripathi A, Pandey N (2018) Does impact of price endings differ for the non-green and green products? Role of product categories and price levels. J Consum Mark 35(2):143–156
21. Tripathi A, Pandey N (2019) Promotional framing of green products: mediating role of sale proneness. J Promot Manag 25(7):983–1008
22. Criss S, Horhota M, Wiles K, Norton J, Hilaire KJS, Short MA, Blomquist KK (2020) Food cultures and aging: a qualitative study of grandparents' food perceptions and influence of food choice on younger generations. Public Health Nutr 23(2):221–230. https://doi.org/10.1016/j.ijinfomgt.2019.05.023
23. Koehler J, Leonhaeuser IU (2008) Changes in food preferences during aging. Ann Nutr Metab 52(1):15–19
24. Adams DC, Salois MJ (2010) Local versus organic: a turn in consumer preferences and willingness-to-pay. Renew Agric Food Syst 331–341
25. Pandey N, Singh G (2012) Marketing issues in SMEs: cases from India. Pearson Education India
26. Singh G, Pandey N (2018) The determinants of green packaging that influence buyers' willingness to pay a price premium. Australasian Market J (AMJ) 26(3):221–230
27. Zhu Z, Chu F, Dolgui A, Chu C, Zhou W, Piramuthu S (2018) Recent advances and opportunities in sustainable food supply chain: a model-oriented review. Int J Prod Res 56(17):5700–5722
28. Hobbs JE, Young LM (2000) Closer vertical co-ordination in agri-food supply chains: a conceptual framework and some preliminary evidence. Supply Chain Manag Int J 5(3):131–143
29. Aung MM, Chang YS (2014) Traceability in a food supply chain: safety and quality perspectives. Food Control 39:172–184
30. Bosona T, Gebresenbet G (2013) Food traceability as an integral part of logistics management in food and agricultural supply chain. Food Control 33(1):32–48
31. Creswell JW, Poth CN (2016) Qualitative inquiry and research design: choosing among five approaches. Sage publications
32. Goulding C (2005) Grounded theory, ethnography and phenomenology. Eur J Mark 39(3/4):294–308
33. Corbin JM, Strauss A (1990) Grounded theory research: procedures, canons, and evaluative criteria. Qual Sociol 13(1):3–21

Chapter 2
Modeling of Risk Factors in Transportation of Perishable Goods Using ISM Approach

Gajanand Gupta, Satyendra Kumar Sharma, Kuldip Singh Sangwan, and Anil Jindal

Abstract Transportation is most well defined as a catalyst of progress. This holds true especially in the case of transportation of perishable food items. Hence, it is tough to viably protect transient products and confirm extreme freshness during blistering and moist climate. In India, the most basic requirement in development of the nourishment preparing industry is the absence of included cold chain facilities. The objective of this research is to present the readily available facts about transport and storage of perishable foods and outline the risk factors involved in this activity. Through research, various risk factors in transportation have been analyzed using ISM technique.

Keywords Supply chain risk · Perishable goods · Interpretive structural modeling · Transportation

2.1 Introduction

Transportation embodies a significant area of human action but surprisingly it is neglected in every business carried out. Transportation plays a vital role in a nation's economy and makes every social and economic activity feasible. It is most well

---

G. Gupta (✉)
School of Mechanical Engineering, Vellore Institute of Technology, Chennai, Tamil Nadu 600127, India

S. K. Sharma
Department of Management, Birla Institute of Technology and Science Pilani, Pilani 333031, India

K. S. Sangwan
Department of Mechanical Engineering, Birla Institute of Technology and Science Pilani, Pilani 333031, India

A. Jindal
Department of Mechanical Engineering, Giani Zail Singh Campus College of Engineering and Technology, Bathinda, Punjab, India

R. S Mor et al. (eds.), *Operations and Supply Chain Management in the Food Industry*, Lecture Notes in Management and Industrial Engineering,
https://doi.org/10.1007/978-981-16-5555-5_2

defined as a catalyst of progress. This holds true especially in the case of transportation of perishable food items. For the ever growing food industry, the demands for fine cooking and tastes have developed, so have the chances for smaller nourishment organizations. In any case, with this comes the significant test of shipping crisp refrigerated and frozen things carefully to their destinations. As per industry estimations, around 100 million tons of perishable freight move by means of non-refrigerated vehicle, and it is evaluated that each year somewhere in the range of 25% of nourishments delivered go surplus because of absence of appropriate stockpiling practices and transportation units. Nourishment wastage at that point makes another arrangement of ecological issues as it changes over into water wastage and prompts host of ozone depleting substance emanations.

Nourishment items regularly disintegrate, because of expanded travel times and incessant stops to serve providers, during the get procedure. It is tough to viably protect transient products and confirm extreme freshness during blistering and moist climate. In India, the most basic requirement in development of the nourishment preparing industry is the absence of included cold chain facilities. The cool chain circulation is a scheme to keep temperature touchy items in prime condition from the purpose of takeoff till it arrives at its last goal. India's most prominent need starting today is for a successful and financially practical cool chain arrangement that will thoroughly incorporate the supply chains for all items, directly from the production facilities to the consumption facilities, in this manner diminishing physical waste and loss of estimation of perishable products. The objective of this research is to present the readily available facts about transport and storage of perishable foods and outline the risk factors involved in this activity.

2.2 Literature Review

In this section, the articles related to risk factors involved in transportation and vehicle routing issues for perishable products have been reviewed. Given the significance of intake of retailers in order to model the stock of perishable products, the authors presented a series of constant demand, the variable demand over time; dependent demand archives and dependent price. Numerous authors have created stock models for crumbling things with variable end-time consumer requirement utilizing linearly or exponentially reduction in requirement [1]. Tarantilis and Kiranoudis [2] presented an edge tolerating calculation to take care of dissemination issue of milk for a dairy organization in Greece. They concentrated on a steering issue with heterogeneous vehicles. Tarantilis and Kiranoudis [3] settled the circulation of items produced using crisp meat in Greece with an open VRP multi-store. They did not consider the extra requirements mirroring the transient idea of the item. Hsu et al. [4] have established a stochastic VRP model for optimum distribution ways, fleet shipping loads and takeoff times for the dissemination of perishable foods. Their aim was to reduce the expense of cargo forwarding vehicles, just as stock costs, energy and fine for violating the time windows requirements. Osvald and Stirn [5] presented an empirical method for the issue of crisp vegetables dispersion in which the effect of the perishable kind of the products has been measured as a piece of the worldwide conveyance costs. Federgruen et al. [6] are among the pioneers researchers who have coordinated directing and stock distribution in a solitary period issue taking irregular retailer's interest for perishable products. They categorized perishable goods in old and new components. The old products would perish in the present timeframe, while new products would be viewed as fresh for as a minimum a period before getting obsolete. Le et al. [7] displayed a multi-period IRP model for perishable goods with fixed life in which client request was deterministic and unsold products with no worth were disposed of. A plan stream way for this issue was proposed, and a lower bound for the issue was accomplished utilizing segment age.

Contrasted with principle things supply chains, sustenance supply chains are consistently dynamically complex and progressively difficult to regulate considering the way that the sustenance thing is transient and has a short time period of practical ease of use. Since nourishments can be time and temperature touchy in nature, they should be appropriately dealt with as far as gathering, readiness, bundling, transportation and taking care of. Temperature is the most significant factor in drawing out or keeping up the timeframe of realistic usability of perishables. Aung and Chang [8] addressed the simulation and wireless sensor network methods to improve the capacity to characterize an ideal objective temperature for multi-product refrigerated stockpiling. From the literature, a total of 9 risk factors are identified, i.e., discussed in following section as the important risk factors involved in transportation of perishable goods.

2.3 Risk Factors Involved in Transportation of Perishable Goods

From the literature review, following comprehensive list of risk factors involved in transportation of perishable goods have been identified. These factors will serve as the input for this further study and modeling.

(1) Lack of security
(2) Improper holding practices for food products and awaiting shipment or inspection
(3) Improper temperature control
(4) Cross contamination
(5) Improper loading practices, conditions or equipment and sanitation
(6) Accident during transportation
(7) Infrastructure problems
(8) Improper transportation unit design
(9) Incorrect usage of packing material or improper packing of transportation unit.

2.4 Methodology

2.4.1 Interpretive Structure Modeling (ISM)

ISM is an interactive management method used to assist research groups in complex issues. ISM is an approach for distinguishing and a bridging connections among explicit components, which characterize an issue and give a methods by which request can be forced on the intricacy of such components. ISM distinguishes the structure inside an arrangement of related components and give a chance to investigate it from a few perspectives. This approach is interpretive as the judgment of the gathering chooses whether and how the various components are connected. It is basic based on shared relationship; a general structure is extracted from the intricate arrangement of components. It is a demonstrating method, as the particular connections and generally structure are depicted in a digraph model.

2.5 Steps Involved in ISM Methodology

Step 1: The first step in ISM involves identification of relevant variables to this issue from literature.

Step 2: This step involves establishing contextual relationships among all the variables by understanding the influence of each element on the other. Overall, it represents which element leads or follows.

Step 3: Structural Self-Interaction Matrix (SSIM) for the variables involved is developed.

Step 4: A reachability matrix from the SSIM is developed using the method of ISM.

Step 5: Different levels of the identified elements determined using the reachability matrix.

Step 6: The levels determined for all the factors in the previous step used to make a digraph.

Step 7: Final ISM is developed.

2.6 ISM on Risk Factors

Step 1: A total of nine elements for risk factors involved in transportation of perishable goods are identified.

Step 2: Contextual relationship among the identified factors was developed based on the assumptions. Each factor is analyzed, and a contextual relationship of "leads to" or "influences" type is chosen, which means that one factor influences the other. On the basis of this, the contextual relationship among all the factors is developed.

Step 3: Based on contextual relationships discussed, the SSIM is developed and finalized, by considering each of the nine factors with respect to the other factors, considering two factors at a time. SSIM shown is Table 2.1.

Step 4: Using the SSIM, the final reachability matrix is determined and presented in Table 2.2.

Step 5: Reachability, antecedent and intersection set are generated using the reachability matrix and presented in Tables 2.3 and 2.4 showing the final level of elements in ISM.

Step 6: Variable hierarchy diagram is generated and shown in Fig. 2.1.

Step 7: Variable hierarchy is converted into an ISM and presented in Fig. 2.2.

Table 2.1 SSIM

Variables	9	8	7	6	5	4	3	2
1	O	V	X	O	O	O	O	O
2	A	V	X	O	O	X	X	
3	A	V	X	O	O	X		
4	X	X	X	O	V			
5	A	X	V	X				
6	O	X	O					
7	V	V						
8	V							
9								

Source Author

Table 2.2 Final reachability matrix

Variable	1	2	3	4	5	6	7	8	9
1	1	0	0	0	0	0	1	1	0
2	0	1	1	1	0	0	1	1	0
3	0	1	1	1	0	0	1	1	0
4	0	1	1	1	1	0	1	1	1
5	0	0	0	0	1	1	1	1	0
6	0	0	0	0	1	1	0	1	0
7	1	1	1	1	0	0	1	1	1
8	0	0	0	1	1	1	0	1	1
9	0	1	1	1	1	1	0	0	1

Source Author

Table 2.3 Reachability, antecedent and intersection set

Variable	Reachability set	Antecedent set	Intersection set
1	1,7,8	1,7	1,7
2	2,3,4,7,8	2,3,4,7,9	2,3,4,7
3	2,3,4,7,8	2,3,4,7,9	2,3,4,7
4	2,3,4,5,7,8,9	2,3,4,7,8,9	2,3,4,7,8,9
5	5,6,7,8	4,5,6,8,9	5,6,8
6	5,6,8	5,6,8	5,6,8
7	1,2,3,4,7,8,9	1,2,3,4,5,7	1,2,3,4,7
8	4,5,6,8,9	1,2,3,4,5,6,7,8	4,5,6,8
9	2,3,4,5,9	4,7,8,9	4,9

Source Author

Table 2.4 Level of elements in ISM

Variable	Reachability set	Antecedent set	Level
1	1,7,8	1,7	Level 2
2	2,3,4,7,8	2,3,4,7,9	Level 4
3	2,3,4,7,8	2,3,4,7,9	Level 4
4	2,3,4,5,7,8,9	2,3,4,7,8,9	Level 6
5	5,6,7,8	4,5,6,8,9	Level 2
6	5,6,8	5,6,8	Level 1
7	1,2,3,4,7,8,9	1,2,3,4,5,7	Level 6
8	4,5,6,8,9	1,2,3,4,5,6,7,8	Level 3
9	2,3,4,5,9	4,7,8,9	Level 5

Source Author

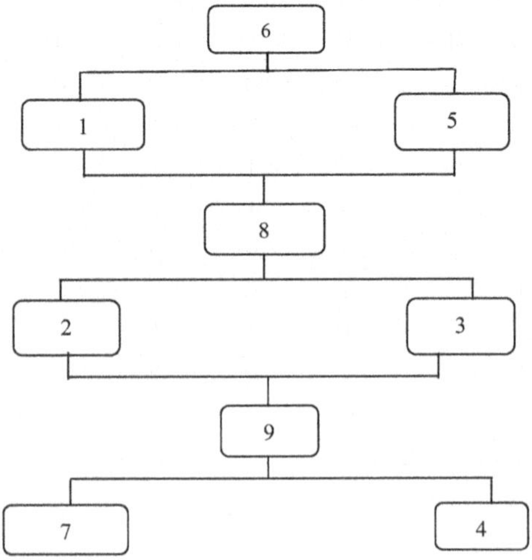

Fig. 2.1 Variable hierarchy diagram. *Source* Author

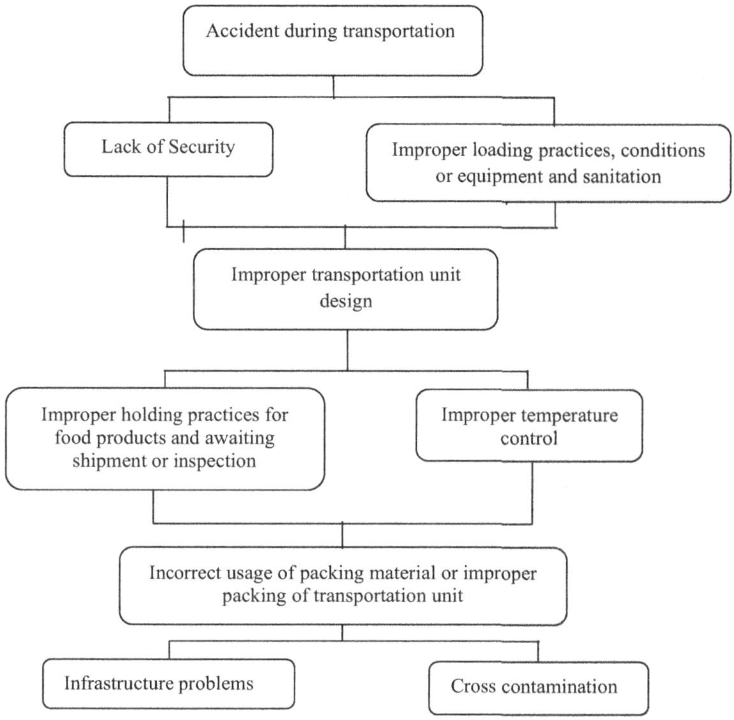

Fig. 2.2 Final ISM. *Source* Author

2.7 Conclusion

With the help of interpretive structure modeling, an ordered directional framework for the issue pertaining to risk factors involved in transportation of perishable goods is developed. This procedure gives us a realistic picture of the situation and the variables (risk factors) involved. This model helps us in determining the key factor related to the problem. The model essentially indicates that accidents is the biggest risk factor in transportation of perishable goods as they occupy Level 1 in the ISM. Moreover, it also indicates the immediate need to cope with the risk associated with accidents in order to avoid loss of perishable products in future. Lack of security and improper loading practices are the tier 2 threats during the transportation process. Lastly, cross-contamination and infrastructure problems occupy the base positions in the mode (Level 6) indicating that they carry minimum risk during transportation.

References

1. Chung KJ, Ting PS (1993) A heuristic for replenishment of deteriorating items with a linear trend in demand. J Oper Res Soc 44(12):1235–1241. https://doi.org/10.1057/jors.1993.202
2. Tarantilis CD, Kiranoudis CT (2001) A meta-heuristic algorithm for the efficient distribution of perishable foods. J Food Eng 50:1–9. https://doi.org/10.1016/S0260-8774(00)00187-4
3. Tarantilis CD, Kiranoudis CT (2002) Distribution of fresh meat. J Food Eng 51:85–91. https://doi.org/10.1016/S0260-8774(01)00040-1
4. Hsu CI, Hung SF, Li HC (2007) Vehicle routing problem with time-windows for perishable food delivery. J Food Eng 80:465–475. https://doi.org/10.1016/j.jfoodeng.2006.05.029
5. Osvald A, Stirn LZ (2008) A vehicle routing algorithm for the distribution of fresh vegetables and similar perishable food. J Food Eng 85 285–195 https://doi.org/10.1016/j.jfoodeng.2007.07.008
6. Federgruen A, Prastacos G, Zipkin PH (1986) Allocation and distribution model for perishable products. Oper Res 34:75–82. https://doi.org/10.1287/opre.34.1.75
7. Le T, Diabat A, Richard JP, Yih Y (2013) A column generation-based heuristic algorithm for an inventory routing problem with perishable goods. Optim Lett 1481–1502. https://doi.org/10.1007/s11590-012-0540-2
8. Aung MM, Chang YS (2014) Temperature management for the quality assurance of a perishable food supply chain. Food Control 40:198–207. https://doi.org/10.1016/j.foodcont.2013.11.016

Chapter 3
Forecast Modelling in Supply Chain Management of Food Processing Industry: A Case Study

P. M. Kulkarni, Sarat Kumar Jena, and Prayag Gokhale

Abstract Supply chain management has grown into one of the most vital elements of the food industry. The main reason for this is, the nature of food which has a short shelf life forecasting the elements of the food supply chain has turned out to be inevitable. The study focuses on the several quantitative and qualitative approaches that are comprehensively adopted for forecasting in the food processing industry; the study also takes account of the scenarios under which each of these approaches can be effectively adopted. In this study, the primary focus is on the methods of forecasting viz. moving averages, exponential smoothening, and the methods allied to the calculation of the error rate in forecasting, like the mean squared error, mean absolute deviation, and mean absolute percentage error are also being illustrated through the use of a case study of Ideal Food Products, Belgaum. According to the calculations, it is found that the error rate for exponential smoothening is comparatively less than that of the moving averages, and it is being observed that the use of multiple methods in forecasting will provide a more accurate judgement to the people involved in the process of food supply chain management.

Keywords Food supply chain · Forecasting · Forecasting error · Moving averages · Exponential smoothening

---

P. M. Kulkarni (✉)
Department of MBA, KLS Gogte Institute of Technology, Udyambag, Belagavi, Karnataka, India
e-mail: pmkulkarni@git.edu

S. K. Jena
Xavier Institute of Management, Xavier Square, Bhubaneswar 751013, India

P. Gokhale
Department of MBA, KLE Dr. M S Sheshgiri College of Engineering and Technology, Udyambag, Belagavi, Karnataka, India

R. S Mor et al. (eds.), *Operations and Supply Chain Management in the Food Industry*, Lecture Notes in Management and Industrial Engineering,
https://doi.org/10.1007/978-981-16-5555-5_3

3.1 Introduction

Prediction of the future happenings is known as forecasting. However, speculation and forecasting are two different terms, wherein speculation is based on personal experience, while forecasting is an application of scientific methods for predicting the future. These future predictions with regards to the business organization include a change in prices of a commodity, demand for production, production forecasting, inventory forecasting, etc., which happens because of uncertain market demand, dynamic business environment, the short life cycle of the product, technological innovation, and globalization. Similarly, a company tries to predict the future demand for its product in the market. These predictions are based on various tools and techniques of forecasting [1].

Forecasting in supply chain management includes planning, evaluation, and decision making with regards to inventory, production, scheduling, design, and distribution of products to retailers, distributors, and consumers. This forecasting is supported by long-term, medium-term, and short-term strategic plans for effective implementation and achievement of higher customer satisfaction [2].

3.1.1 Importance of Forecasting in Supply Chain Management

Forecasting plays a significant role in several areas of the enterprise, especially for supply chain decisions. For instance, in the case of procurement of raw materials, the supply chain supports in forecasting and procurement of raw material for production. Therefore, forecasting in supply chain management is important for the organization.

The importance of forecasting in supply chain management is as follows;

1. Customer's needs and demands change; hence, it is important to understand these changing needs and forecasted inventory for the production [3].
2. Inventory for production includes durable and non-durable products. The forecasting technique supports the identification of the right quality of inventory required for production and reduced the cost of inventory storage in the organization [4].
3. Recent technology is influencing supply chain management. These technologies include artificial intelligence, blockchain, and big data analytics. These technologies support the inventory forecasting for production [5, 6].

3.1.2 Challenges of Forecasting in Supply Chain Management

The main objective of supply chain management is to achieve a smooth flow of inventory to achieve a higher level of production. This is possible through the implementation of best practices concerning demand forecasting, lead times, and stock quantities. However, there are challenges with regards to the implementation of best practices as follows;

1. Forecasting is influenced by demand, changing consumption patterns of consumers, and the influence of the unorganized competitive market. Therefore, attaining accuracy in forecasting is a challenge to the organization [7].
2. Companies invest heavily in marketing and communication to attract consumers and retain consumers; this influences supply and demand in the market. Hence, the marketing and sales strategies of competitors influence the forecasting of the inventory [8].
3. Storage of raw material and storage of final finished products need effective collaboration with the inventory department and production department. In the event of a lack of collaboration and coordination with these departments, accuracy of forecasting is reduced and influences the production process [9].
4. Forecasting is greatly influenced by the effective strategy of supply chain management, especially for the food industry, as they operate in perishable products. Food manufacturing companies need to develop strategies that are in coordination with organizational strategy [10].
5. Supply chain strategy includes food products, manufacturing or processing, vendor management, and logistics management. All these factors need to be included in the supply chain management strategy of a food company [11].
6. Strategic supply chain management needs to be supported with the latest technology which includes artificial intelligence and the Internet of things; these technologies support in providing real-time data with regards to products logistics, vendor sales, and distributions, and warehousing, which is essential for effective supply chain management strategic decisions [12].

3.1.3 Supply Chain Forecasting Methods

Forecasting methods in supply chain management are influenced by several factors such as seasonal demand for food inventory, labour cost, change in consumer buying habits, and distribution and logistics challenges. Selection of the right method for supply chain forecasting is important. There are various methods of forecasting, these methods are presented in Table 3.1.

Table 3.1 Methods and selection of supply chain forecasting in the food industry

S. No.	Method of forecasting	Application of forecasting
1	**Delphi Method**: It is a process to arrive at an opinion or decision through survey of panel of experts. Opinion and decision are collected through a questionnaire and shared with the panel of expert after each round [13]	Long-term forecasting which includes food product demand, food products sales forecasting, and new food product demand analysis
2	**Market Research**: This method of forecasting identifies new markets, new products by contacting the future customers for the company. This method provides information with regards to perception of the customers towards the products and services [13]	Long-term forecasting which includes identification of new areas of food ingredients for food production, changing supply chain channels, future demand for food product, consumer preferences
3	**Panel Consensus**: In this method, internal experts from cross-functional areas are involved and discussion related to services and activities feedback is collected, and finally, they reach a conclusion through discussion [13]	Forecasting of long-range and new products, and sales forecasting
4	**Visionary Forecast**: Future consequence of the event is analysed through data and forecasting is undertaken in the company [13]	Forecasting future logistic management, distribution channels, sales of food products, new market for food products, changing pattern of consumers
5	**Historical Analogy**: Past sales related to battery of products and services are considered for forecasting of products and services [13]	Forecasting of long-range and new products, and sales forecasting
6	**Moving Average**: In this process, average price is considered for a period of time, days. This is applied to understand the average price for the product or service [13]	Inventory control for low-volume items
7	**Exponential Smoothing**: In this tool, exponential method is applied and an average data of price is collected for forecasting [13]	Production, inventory control, forecasts of margins and financial data
8	**Box–Jenkins**: It is a mathematical technique for forecasting based from a range of time duration. The technique of autoregressive integrated moving average (ARIMA) is related to this technique [13]	Production and inventory control for large volume its
9	**X-11**: In this method time series in estimation of trends with regards to seasonal variation is provided by applying right weightage for end results of the forecasting [13]	Logistics, department forecasts of company sales

(continued)

Table 3.1 (continued)

S. No.	Method of forecasting	Application of forecasting
10	**Trend Projection**: In this method, huge data is used for forecasting, the data related to past information with regards to sales or production is applied for forecasting [13]	New-product forecasts (particularly intermediate and long term)
11	**Regression Model**: In this method, dependent and independent factors are considered for forecasting is applied to understand the factors influencing the forecasting [13]	Sales by products, inventory by class, margins
12	**Econometric Model**: In this technique, forecasting related to economic factors is considered to project the future demand for products and services in the organization [13]	Sales by products, inventory by class, margins
13	**Intention to Buy**: New products and services sales are forecasted through this method. In this method, likely intention of buying is forecasted through this method [13]	Sales by products, inventory by class, margins
14	**Economic Input–Output Model**: Influence of economic factors is analysed and forecasted through this method; the results are positive and negative with regard to influence on economic aspects of business [13]	Industry sector sales, company sales
15	**Diffusion Index**: This is an index which is applied to understand the common trend through a mathematical technique [13]	Forecasting of sales by inventory class
16	**Leading Indicator**: Business factors and influence on indictors are predicted and forecasted through this technique [13]	Forecasting of sales by inventory class
17	**Life-Cycle Analysis**: It is a tool which considers production to sales of a products and forecasting is undertaken through this method [13]	Forecasting of new product sales

Source Author

3.1.4 Forecasting Duration And Its Application Models

Modelling for forecasting is influenced by time duration that is short-term, medium-term, and long-term forecasting. There are different modelling methods of forecasting based on the time duration; the details with regards to methods based on time duration are presented in Table 3.2.

Table 3.2 Time duration and modelling of forecasting[14]

Application	Short term (0–3 months)	Medium term (3 months–2 years)	Long term (more than 2 years)
Forecast quality	Individual products and services	Sales forecasting	Total sales
Decision area	Inventory forecasting Production scheduling Logistic forecasting	Distribution forecasting Production planning Inventory purchase planning	Inventory capacity planning Storage location and planning
Forecasting models	• Moving average • Exponential smoothing • Box–Jenkins • X-11 • Trend projections • Regression model • Econometric model • Intention to buy and anticipations surveys • Economic input–output model • Diffusion index • Leading indicator • Life-cycle analysis	• Exponential smoothing • Delphi method • Market research • Panel consensus • Visionary forecast • Historical analogy • Regression model • Econometric model • Intention to buy and anticipations surveys • Economic input–output Model • Diffusion index • Leading indicator • Life-cycle analysis judgement	• Delphi method • Market research • Panel consensus • Visionary forecast • Historical analogy • Regression model • Econometric model • Intention to buy and anticipations surveys • Economic input–output model • Diffusion index • Leading indicator • Life-cycle analysis judgement

3.1.5 Models of Forecasting and Levels of Accuracy in Forecasting

Forecasting models are influenced by the level of accuracy that means, any forecasting has a component of error. Therefore, while understanding the forecasting, the error level needs to be reduced to reach actual forecasting; there are three most popular methods of understanding the errors in forecasting [15]. These three methods of forecasting error are (1) mean squared error (MSE), (2) mean absolute deviation (MAD), and (3) mean absolute percentage error (MAPE) [15].

3.1.5.1 Mean Squared Error (MSE)

In this method of forecasting error, the standard deviation of forecast error is considered while forecasting. The formula for calculating mean squared error [16].

$$\text{MSE} = \frac{1}{N}\sum_{t=1}^{T}(y_t - f_t)^2 = \frac{\text{SSE}}{N} \tag{3.1}$$

where

y_t = is the actual outcome value at time t.
f_t = is the forecast value at time t.
SSE = is the sum of squared errors.
N = No. of observations.

3.1.5.2 Mean Absolute Deviation (MAD)

It is a set of datasets that shows the average distance between each data point and the mean value. This method shows the variability in a dataset. The formula for calculating mean absolute deviation (MAD) is as under [14]

$$\text{MAD} = \sum_{i=1}^{n} |e_i - \bar{e}| \tag{3.2}$$

where

- e_i = performance value for period i
- e = average Value
- n = number of observations.

3.1.5.3 Mean Absolute Percentage Error (MAPE)

This is a measure to forecast the accuracy through percentage. The MAPE is calculated by taking an average absolute per cent error minus the real value which is divided by the real value. The formula for mean absolute percentage error is as under;

$$\text{MAPE} = \frac{1}{n} \sum_{i}^{n} =|\frac{v(t) - P(t)}{V(t)} * 100 \tag{3.3}$$

where $V(t)$ = Actual value, $P(t)$ = Forecast value and n = Number of ex-post periods.

3.2 Case Study on the Effective Forecasting of Supply Chain Decisions of Ideal Food Products, Belgaum, Karnataka-India

3.2.1 Background of the Company

Ideal Food Products established in the year 1950 to manufacture and sell edible syrups and fruit pulps. At the opening of this unit, there were very few companies manufacturing these two products. The location of the unit was an added advantage to Ideal Food Products, as they were able to cater to the needs of customers in the region of North Karnataka, Goa, and South Maharashtra.

3.2.2 Problem Statements

The sales of the company have shown an increase and decrease in the last decade due to the entry of new companies and changing consumer needs and demands. Further, there was pressure on the inventory cost due to an increase in the price of inventory, especially specific key fruits such as mango, strawberry, and dry fruits. In the last few years, the cost of logistics and storage has increased by 15%. The challenges with regards to Ideal Food Products are as follows;

1. The total income in April 2019 was 150,000 and in March 2020; it is reduced to 141,000. Hence, there was a reduction of 9000 income reduction from April 2019 to March 2020(Annexure 1)

   The production cost has increased by 39,500 in April 2019 to 52,900 in March 2020. However, the cost of raw material has shown fluctuation in cost due to the influence of seasonal demand for fruits.
2. Factors that have influenced the supply chain management shows that the average inventory cost is 42,500; this has led to an increase in the production cost to 43,000. Therefore, inventory cost has influenced the overall production cost of Ideal Food Products. (Annexure 2)

3.2.3 Forecasting Modelling for Ideal Food Products

The following forecasting models are used to understand sales forecasting and inventory cost forecasting. The moving average method is applied to understand the forecasting of sales. While exponential smoothing modelling is applied to forecast the inventory cost of Ideal Food Products

3.2.4 Calculations of Forecasting

In this section, the study presents the results with regards to two models of forecasting; they are moving average forecasting and exponential smoothing forecasting.

Table 3.3 Three-month moving average forecasting

Period	Sales (Rs. '000)	Forecasts (Rs. '000)
July	150	–
August	120	–
September	151	–
October	95	(150 + 120 + 151)/3 = 140.3333
November	99	(120 + 151 + 95)/3 = 122
December	112	(151 + 95 + 99)/3 = 115
January	102	(95 + 99 + 112)/3 = 102
February	98	(99 + 112 + 102)/3 = 104.3333
March	95	(112 + 102 + 98)/3 = 104
April	–	(102 + 98 + 95)/3 = 98.3333

Source Author

3.2.4.1 Sales Forecasting Through Moving Average Model of Forecasting

Moving average is calculated as under

$$\hat{T}t = 1mk\Sigma j = -kyt + j \tag{3.4}$$

where $m = 2k + 1$, i.e. t is obtained by calculating the average of values in the time series within k periods of t. The observations are likely to be near in value which are proximate in time. Therefore, the randomness in the data is eliminated by the averages, leading to a smooth trend cycle. It is called as moving average of an order m. The calculations with regards to the sales of Idea Food Products are presented in Tables 3.3 and 3.4 depicts the error metrics for moving average forecasting.

Therefore, the forecast for April using a three-month moving average is 98.3. The MAD (mean average deviation) = 14.44, the MSE (mean square error) = 452.37, and the MAPE (mean average percentage error) = 14.93%.

3.2.4.2 Inventory Forecasting Through Exponential Smoothing Model of Forecasting

The cost of higher inventory influences the profitability of the company; to understand the influence of inventory cost, an exponential smoothing model of forecasting is applied in the study as shown in Table 3.5, whereas Table 3.6 shows the error metrics.

The exponential smoothing is calculated as follows

$$\hat{y}T + h|T = yT, \text{ for } h = 1, 2, \ldots \tag{3.5}$$

Table 3.4 Calculations of the corresponding error metrics for moving average forecasting

Period	Sales (Rs. '000)	Forecasts (Rs. '000)	Abs. error	Error^2	Abs. % error
July	150	–			
August	120	–			
September	151	–			
October	95	140.3	|95 – 140.3333| = 45.3333	(95 – 140.3333)2 = 2055.11	|95 – 140.3333|/95 = 47.72%
November	99	122	|99 – 122| = 23	(99 – 122)2 = 529	|99 – 122|/99 = 23.23%
December	112	115	|112 – 115| = 3	(112 – 115)2 = 9	|112 – 115|/112 = 2.68%
January	102	102	|102 – 102| = 0	(102 – 102)2 = 0	|102–102|/02 = 0%
February	98	104.3	|98 – 104.3333| = 6.3333	(98 – 104.3333)2 = 40.11	|98 – 140|/98 = 6.46%
March	95	104	|95 – 104| = 9	(95 – 104)2 = 81	|95 – 140|/95 = 9.47%
April	–	98.3			

Source Author

Table 3.5 Exponential smoothing (ES) forecast with smoothing constant $\alpha = 1$

Period	Inventory (Rs. '000)	Forecasting (Rs. '000)
January	45	45
February	38	45 + 1·(45 – 45) = 45
March	56	45 + 1·(38 – 45) = 38
April	35	38 + 1·(56 – 38) = 56
May	39	56 + 1·(35 – 56) = 35
June	45	35 + 1·(39 – 35) = 39
July	42	39 + 1·(45 – 39) = 45
August	41	45 + 1·(42 – 45) = 42
September	39	42 + 1·(41 – 42) = 41
October	–	41 + 1·(39 – 41) = 39

Source Author

Table 3.6 Exponential smoothing (ES) forecast corresponding error metrics

Period	Inventory (Rs. '000)	Forecasting (Rs. '000)	Abs. error	Error^2	Abs. % error
January	45	45	|45 − 45| = 0	(45 − 45)2 = 0	|45 − 45|/45 = 0%
February	38	45	|38 − 45| = 7	(38 − 45)2 = 49	|38 − 45|/38 = 18.42%
March	56	38	|56 − 38| = 18	(56 − 38)2 = 324	|56 − 38|/56 = 32.14%
April	35	56	|35 − 56| = 21	(35 − 56)2 = 441	|35 − 56|/35 = 60%
May	39	35	|39 − 35| = 4	(39 − 35)2 = 16	|39 − 35|/39 = 10.26%
June	45	39	|45 − 39| = 6	(45 − 39)2 = 36	|45 − 39|/45 = 13.33%
July	42	45	|42 − 45| = 3	(42 − 45)2 = 9	|42 − 45|/42 = 7.14%
August	41	42	|41 − 42| = 1	(41 − 42)2 = 1	|41 − 42|/41 = 2.44%
September	39	41	|39 − 41| = 2	(39 − 41)2 = 4	|39 − 41|/39 = 5.13%
October	–	39			

Source Author

Hence, the raw technique adopts that the most current observed values are the only significant, and all earlier interpretations offer no evidence for the future. This can be understood as a weighted average in which all the weightage is set to the terminal observed value. Using the average technique, all future forecasts are equal to a simple average of the experimental data.

$$\hat{y}T + h|T = 1TT\Sigma t = 1yt, \text{ for } h = 1, 2, \ldots \tag{3.6}$$

So, the average technique accepts that all observed values are of equal significance and provides these observed values equal weightage while making a forecast. We time and again need something amid these two limits. For instance, it might be practical to assign higher weightages to most current observed values than the values observed from the past. This exactly is the context governing the simple exponential smoothing. The forecasts are made through weighted averages, in which the weightages will reduce exponentially as the observed values originate from further in the past—the minimum weights are related to the oldest observed values:

$$\hat{y}T + 1|T = \alpha yT + \alpha(1 - \alpha)yT - 1 + \alpha(1 - \alpha)2yT - 2, \ldots \tag{3.7}$$

Table 3.7 Application forecasting modelling and level of accuracy for forecasting

Application	Short term (0–3 months)	Medium term (3 months–1 year)
Forecast quality	Food products sales	Inventory cost
Decision area	Sales forecasting	Sales forecasting
Forecasting models	Moving average	Exponential smoothing
Forecasting error	MAD = 14.44 MSE = 452.37 MAPE = 14.93%	MAD = 6.89, MSE = 97.78 MAPE = 16.54%

Source Author

where $0 \leq \alpha \leq 1$ is the smoothing parameter. $T + 1$ is the weighted average of all of the observed values in the progressions $y1, \ldots, yT$. The rate at which the weights drop is controlled by the parameter Alpha (α).

Therefore, for the given time series data, the forecast for October is 393939 using an exponential smoothing technique with $\alpha = 1$. MAD = 6.89, MSE = 97.78, and MAPE = 16.54%

3.2.5 Outcome of the Forecasting Modelling for Ideal Food Products

In the above case study, different models of forecasting were adopted. Sales and inventory cost were considered an important construct for forecasting. Therefore, considering the case situation, moving average and exponential smoothing modelling were applied for the study for short term forecasting and medium-term forecasting. The results and outcomes are presented in Table 3.7

The forecasting modelling indicated that Ideal Food Products needs to use measures to reduce distribution costs by identifying areas of cost such as transport cost, labour cost, and application of technology, which would improve these costs. Finally, sales of the company need to be improved by identifying new markets and identify new channels of distribution such as online marketing websites, e-retailers, which would provide additional revenue to the company.

3.3 Conclusions

The increasing rate of inflation and its pressure on the supply chain elements has led the enterprises to strategically adopt the best practices in the supply chain and to forecast and optimally reduce the cost on the elements of the supply chain and increase its profitability.

The study presented various forecasting tools that can be adopted by the enterprises depending on the type of product, the duration of projection, and the accuracy that the firm accepts from such projections. The various methods as suggested in this study can be adopted by the enterprises as per their needs and requirements, and it is always beneficial to employ more than one method of forecasting for generating more dependable results.

The best supply chain practices with regards to the food industry include demand forecasting and collaborative planning, forecasting, and replenishment. In this chapter, we have focused on forecasting the elements of supply chain decisions with special reference to the food processing industry.

It is being put forth from the study that the strategic decisions concerning supply chain decisions in the food processing industry need to be accurate, driven by real-time data for effective strategic implementation, and match the customer's expectations to achieve a higher level of customer satisfaction.

The study covers various quantitative and qualitative methods of forecasting that are extensively used in the food processing industry, along with the scenarios under which each of the methods can be effectively adopted for forecasting. It also focuses on the accuracy of the results through the various techniques and discusses the mean absolute deviation which is the average deviation between the actual and the estimated results and the cumulative error is the summation of all the errors.

Lastly, the case study on the Ideal Food Products benefits in understanding the application of various techniques of forecasting in the food processing sector. The illustrations, shown through company data, related to the moving averages and exponential smoothening models provide a better insight into these methods in forecasting.

3.4 Disclosure

Conflict of Interest Dr. P. M. Kulkanri and co-authors declare that there is no conflict of interest and have not received any funds from the company under study and the company under the study has issued required permissions for publication of data and information in the book chapter.

Annexure 1—Income Statement of Ideal Food Products (*Source* Author)

Particulars	Apr 19	May 19	June 19	July 19	Aug 19	Sept 19	Oct 19	Nov 19	Dec 19	Jan 20	Feb 20	Mar 20
Income												
Sales												
Cash sales	112.50	90.00	113.25	71.25	74.25	84.00	76.50	73.50	71.25	109.50	68.25	89.25
Credit sales	37.50	30.00	37.75	23.75	24.75	28.00	25.50	24.50	23.75	36.50	22.75	29.75
Total income (A)	150.00	120.00	151.00	95.00	99.00	112.00	102.00	98.00	95.00	146.00	91.00	119.00
Expenditure												
Raw materials												
Purchase cost	36.00	30.40	44.80	28.00	31.20	36.00	33.60	32.80	31.20	41.60	30.40	35.20
Inventory cost	9.00	7.60	11.20	7.00	7.80	9.00	8.40	8.20	7.80	10.40	7.60	8.80
Total cost of raw materials	45.00	38.00	56.00	35.00	39.00	45.00	42.00	41.00	39.00	52.00	38.00	44.00
Other expenditure												
Storage cost	4.80	2.20	5.10	2.80	2.70	4.80	4.70	3.90	3.70	5.40	2.20	4.60
Production cost	33.00	38.00	55.00	41.00	48.00	43.00	46.00	39.00	41.00	49.00	42.00	49.00
Distribution cost	1.20	1.30	1.80	1.50	1.70	1.60	1.50	1.40	1.50	1.30	1.50	1.40
Marketing cost	0.50	0.80	1.10	1.30	1.60	1.90	2.20	2.80	2.40	3.10	3.50	3.90
Cost of production	39.50	42.30	63.00	46.60	54.00	51.30	54.40	47.10	48.60	58.80	49.20	58.90
Total cost (B)	84.50	80.30	119.00	81.60	93.00	96.30	96.40	88.10	87.60	110.80	87.20	102.90
Profit before interest and tax ($A - B$)	65.50	39.70	32.00	13.40	6.00	15.70	5.60	9.90	7.40	35.20	3.80	16.10
Tax (30%)	19.65	11.91	9.6	4.02	1.8	4.71	1.68	2.97	2.22	10.56	1.14	4.83
Profit after tax	45.85	27.79	22.40	9.38	4.20	10.99	3.92	6.93	5.18	24.64	2.66	11.27

Annexure 2—Supply Chain Management Cost of Ideal Food Products (*Source* Author)

Months	Sales (Rs. '000)	Inventory cost (Rs. '000)	Storage cost (Rs. '000)	Production cost (Rs. '000)	Distribution cost (Rs. '000)	Marketing cost (Rs. '000)	Total cost of production (Rs. '000)
Apr-19	150.00	45.00	4.80	33.00	1.20	0.50	84.50
May-19	120.00	38.00	2.20	38.00	1.30	0.80	80.30
Jun-19	151.00	56.00	5.10	55.00	1.80	1.10	119.00
Jul-19	95.00	35.00	2.80	41.00	1.50	1.30	81.60
Aug-19	99.00	39.00	2.70	48.00	1.70	1.60	93.00
Sep-19	112.00	45.00	4.80	43.00	1.60	1.90	96.30
Oct-19	102.00	42.00	4.70	46.00	1.50	2.20	96.40
Nov-19	98.00	41.00	3.90	39.00	1.40	2.80	88.10
Dec-19	95.00	39.00	3.70	41.00	1.50	2.40	87.60
Jan-20	146.00	52.00	5.40	49.00	1.30	3.10	110.80
Feb-20	91.00	38.00	2.20	42.00	1.50	3.50	87.20
Mar-20	119.00	44.00	4.60	49.00	1.40	3.90	102.90

Annexure 3—Sales Forecast of Ideal Food Products

Year	Apr 19	May 19	June 19	July 19	Aug 19	Sept 19	Oct 19	Nov 19	Dec 19	Jan 20	Feb 20	Mar 20	Apr 20	May 20	June 20
Sales (Rs. '000)	150.00	120.00	151.00	95.00	99.00	112.00	102.00	98.00	95.00	146.00	91.00	119.00	141.00	110.38	114.77

References

1. Korpela J, Tuominen M (1996) Inventory forecasting with a multiple criteria decision tool. Int J Prod Econ 45(1–3):159–168
2. Zhong R, Xu X, Wang L (2017) Food supply chain management: systems, implementations, and future research. Ind Manag Data Syst 117(9):2085–2114
3. Gružauskas V, Gimžauskienė E, Navickas V (2019) Forecasting accuracy influence on logistics clusters activities: the case of the food industry. J Clean Prod 240:118225
4. Ilbery B, Maye D, Kneafsey M, Jenkins T, Walkley C (2004) Forecasting food supply chain developments in lagging rural regions: evidence from the UK. J Rural Stud 20:331–344
5. Behnke K, Janssen MFWHA (2020) Boundary conditions for traceability in food supply chains using blockchain technology. Int J Inf Manag 52:101969
6. Luo J, Ji C, Qiu C, Jia F (2018) Agri-food supply chain management: bibliometric and content analyses. Sustainability 10(5):1573
7. Luthra S, Mangla SK (2018) When strategies matter: adoption of sustainable supply chain management practices in an emerging economy's context. Resour Conserv Recycl 138:194–206
8. Dellino G, Laudadio T, Mari R, Mastronardi N, Meloni C (2018) A reliable decision support system for fresh food supply chain management. Int J Prod Res 56(4):1458–1485
9. Priyadarshi R, Panigrahi A, Routroy S, Garg GK (2019) Demand forecasting at retail stage for selected vegetables: a performance analysis. J Model Manag 14(4):1042–1063
10. Emamisaleh K, Rahmani K (2017) Sustainable supply chain in food industries: drivers and strategic sustainability orientation. Cogent Bus Manag 4(1):1345296
11. Vodenicharova MS (2020) Supply chain study in food industry in Bulgaria. Int J Retail Distrib Manag 48(9):921–938
12. Kumar V, Chibuzo EN, Garza-Reyes JA, Kumari A, Rocha-Lona L, Lopez-Torres GC (2017) The impact of supply chain integration on performance: evidence from the UK food sector. Procedia Manuf 11:814–821
13. Chambers JC, Mullick SK, Smith DD (1971) How to choose the right forecasting technique. Harvard University, Graduate School of Business Administration
14. Pham-Gia T, Hung TL (2001) The mean and median absolute deviations. Math Comput Model 34(7–8):921–936
15. Mahmoud E (1984) Accuracy in forecasting: a survey. J Forecast 3(2):139–159
16. Hyndman RJ, Koehler AB (2006) Another look at measures of forecast accuracy. Int J Forecast 679–688

Chapter 4
Cost Competitiveness Through Value Chain Analysis: A Study About Black Pepper

Avanish Shandilya, Disha Gulia, and Sapna Arora

Abstract Black pepper has long been an important commercial spice in many parts of the world. In recent times, the surplus of supply and rising proportion of spice imports have harmed prices in major markets. India is a good example of this, where farmers have been offering their produce at exceptionally low prices. This study covers Sulthan Bathery, Kalpetta, Mattancherry, Ernakulam, and Kochi of Kerala and Gudaloor of Tamil Nadu to identify the procurement and value chain models of these markets. Identification and mapping of stages involved, intermediaries' functions, and value chain mapping thus lay the foundation for this chapter. The supply chain models shown in the chapter help to understand how the supply chain looks like and what are the nodes the commodity touches while in the value addition process. Also, the cost sheet was studied to understand the financing of the value chain of black pepper. The cost sheets will help an entrepreneur in coming to a fair deal with the traders during the procurements. When it comes to choosing a procurement market, the proprietor should understand its capability and standards to satisfy the goals of the enterprise.

Keywords Cost sheet · Value chain · Black pepper · Procurement · Wholesaler

4.1 Introduction

The spices production in India has increased tremendously over a period of time. The matter of fact is that 75 varieties of the 109 identified by International Standard Organization (ISO) are produced in India. India is the world's largest producer and exporter of spices, contributing for half of global spice trade and exporting more than 60% of its output [1]. Spices and seasoning mixes market were net valued at Rs. 40k in 2017. The black pepper is designated as "King of Spices" due to its anti-oxidant and nutrient property and is used for flavoring the food because of pungent

A. Shandilya · D. Gulia · S. Arora (✉)
Department of Food Business Management and Entrepreneurship Development (FBMED), National Institute of Food Technology Entrepreneurship and Management (NIFTEM), Kundli, Sonipat, Haryana 131028, India

R. S Mor et al. (eds.), *Operations and Supply Chain Management in the Food Industry*, Lecture Notes in Management and Industrial Engineering,
https://doi.org/10.1007/978-981-16-5555-5_4

hot tasting spice as well as medicinal to cure stomach ache, bronchitis, and tumors and is commonly used as a pain reliever. The market is dominated by many domestic masala companies which include many local and regional players. The increasing demand for the pepper is providing new market opportunities for the vendors to enter with a hope to gain more margin as prices are continuously increasing. Even though the production of black pepper is not increasing from last few years, yet Indian black pepper is available at the highest prices [2]. As per the current market scenario, the new crop of black pepper is expected to account for about 30–35% of the market. Retail, commercial, and grocery or kirana shops are the three general categories in which the industry can be divided. The quick service restaurant (*QSR*) segment has also grown at a very good pace over a decade and is still growing at a very decent pace. The Indian QSR market was valued at Rs. 20k Crores in 2016 and is expected to reach Rs. 50k Crores by 2021. There are a lot of opportunities for young entrepreneurs in India to start a venture in the spices and seasonings segment.

4.1.1 Black Pepper Process

Black pepper is grown in the Western Ghats [1] covering Kerala, Karnataka, and Tamil Nadu. The crop is grown, harvested, and preprocessed at the farm level. The berries are washed and cleaned post-harvesting and then dried in shade for 48 h or blanched and then dried. The blanched berries have a very less microbial load, and hence, the berries are not contaminated and safe from contamination. The shade-dried berries give the highest volatile oil yield. A standardized process of blanching to make good quality black pepper was given by CFTRI, Mysore. The process is simple and can be easily done by the farmers at the farm level, which will cut down freight, labor, and other resources.

The freshly harvested berries are plucked from the spikes and cleaned. After cleaning, the berries are passed into a perforated aluminum trough and dipped in a vessel of boiling water for almost a minute. Berries are thereafter drained and poured onto a concrete floor to dry in sun. Post this process, the berries are packaged and transported from the farm to the collectors.

The collectors collect the berries from the farmers and do primary pre-cleaning and drying (if required). The collected lots of berries are then sold to the suppliers or manufacturers who procure based on the customer's requirements.

4.1.2 Procurement

The procurement of black pepper is not easy as it seems. The market for the commodity is very volatile and shows ups and downs due to many factors. Earlier, the pepper market in India was concentrated in the southern part of the country but now black pepper is available all across the country whether one wants to buy cheaper Vietnamese low-quality pepper or superior quality Idduki pepper; the commodity is available everywhere, increasing the concerns for adulteration, mixing, black marketing and many more. The Vietnamese black pepper was routed to India via. Sri Lanka as the import duty if a trader in India imports from Sri Lanka is far less than the import duty on Vietnam imports,

Moreover, the SAFTA agreement of SAARC countries had supported the traders in India to procure black pepper from Sri Lanka at 8% import duty instead of paying 54% from Vietnam. In case of demand up to 2500 tons, need not to pay any import duty to Sri Lanka. This has created turmoil among the Indian pepper growers slowly; India will become a net importer of black pepper. Recently, the government enforced a minimum import price of Rs. 500 for black pepper imports irrespective of the source to curb the misuse of the SAARC treaty and assured farmers to continue production of the crop as the growers had no choice but to stop cultivating the black pepper.

The present chapter deals with the procurement of raw unground spices by the industrial sector. With the view that the spices market will attain very high growth soon, a lot of big players of FMCG and food segments like HUL, ITC, TATA, NESTLE, McCormick, etc. are trying to or has entered the market hitting the B2B and retails section. As is evident from market research conducted by P&S Market, research showing that the spices and seasonings market may attain $30,412.8 in 2023 growing at a growth rate of 6% during the period of the forecast (2018–2023).

4.1.3 Meaning of Value Chain

The term "value chain" describes the entire set of tasks that are essential for bringing a product from conception to completion (which requires a mix of physical and digital processes), delivery to final customers, and final transition and feedback from various producer services). Food value chains are created to gain a competitive advantage by collaborating in a business venture. Producers, manufacturers, distributors, food service providers, retailers, and supporting organizations are all linked by the food value chain [3].

4.2 Review of Literature

The crop "black pepper" is one of the high-value spices, and the quality of Indian spices is also superior in comparison to other markets. The authors reviewed the various articles available on the value chain or supply chain of black pepper, however, found very few relevant studies which are summarized here:

Sharangi and Acharaya [4] studied supply chain of the chosen orchard and plantation crops such as banana and spices. The people's participation in value addition after post-harvest processing was also analyzed for the same crops. Hema et al.

[5] conducted research to determine the factors that influence its production, assess farmer profitability, and investigate the price behavior of black pepper. The increasing input cost and depressed prices in domestic or global markets discouraged farmers to continue the production of black pepper [6]. Over the years, the domestic and export prices have moved at lockstep and prices have appeared to reach equilibrium sooner in the long term than before liberalization. Later, Paulrajan [7] looked at the routes and distances spices traveled from the field to the market in both conventional and coordinated retailing.

Prabhavathi et al. [8] discussed the practices involved, supply chain models and their nodes, the price spread across the supply chain, and the efficiencies of the supply chain models. The work provides a theoretical percentage of the price spread across the supply chain for the red chili. Further, Yogesh et al. [9] have also conducted a study on black pepper production and export performance. Ganpathy et al. [10] identified that the study of manufacturing methods and quality control at the farm level, as well as the evaluation of the black pepper value chain, are not different from, but rather a component of value chain analysis. Value addition takes place at each stage of the supply chain process starting from production, processing, storage, and finally selling to the end consumers. Dessie et al. [11] studied the red pepper value chain to determine the factors for supplying to the Abshire Woreda market. The study suggests a measure to increase the producer share through an active role of cooperatives and resulting in more production for the red pepper. Majo and Elsa [12] identified that the Spice Board of India must amp up the promotion and development activities at the grass-root level to help the small farmers engaged in the production of cardamom. It must focus on creating new strategies to increase demand potential and prospects, to assist cardamom growers and exporters in maintaining market linkages, and to expand both domestic and foreign markets by increasing consumer awareness and assisting farmers in producing high-quality Cardamom. Further, Cristiano Mandinga Bonfim et al. [13] pointed out that the local agribusiness for black pepper is still in its infancy. They proposed pepper seller certification, a collaboration among independent producers who are unable to join the cooperative, a public–private partnership to establish structured trade route with neighboring countries, and a program to improve supervisory skills among small traders. Krishnaraj et al. [14] studied the existing logistical route of the spice trade to find food mileage for both the traditional and organized retailers. They concluded that, as a result of logistical changes, nutritional miles for coordinated and organized stores have risen fairly. The major factor responsible for higher food miles' for the organized retailer is the availability of more supply sources nearby the harvesting areas that are away from retail centers. Friesenbichler et al. [15] explored how value chain integration affects the price fluctuations at the sector level for European Union. Producer price inflation is lower in sectors that become more backward integrated. Forward integration in the supply chain is weaker as compared to the backward integration for these sectors.

The intended work gives an overview for new entrepreneurs to understand the supply chain model, procurement strategies, and procurement cost of high valued spices namely black pepper. The cost sheets are made considering in-depth processes involved in the supply chain. It was inevitable to make the cost sheets without understanding the processes involved, and hence, the cost sheets have been made in synchronization with the processes involved.

4.3 Problem Formulation

4.3.1 Objectives

The chapter aims to develop the cost sheets related to the processing and manufacturing of black pepper and to understand the strategies involved in maintaining the supply chain and procurement decisions. The rationale of the study is to look at each aspect of cost to achieve cost competitiveness even if the procurement cost could be minimized with little change.

4.3.2 Research Methodology

The present study is exploratory in nature. The primary data was collected through structured questionnaires prepared from farmers, suppliers, and brokers involved in the supply chain of the black pepper crops. The sample units include the markets surveyed and investigated under the project. The number of persons interviewed for the project is 23 of which 16 are wholesale traders and exporters, 4 cultivators, and 3 employees of Spices Board of India.

4.3.3 Data Collection

The following markets have been selected for data collection: Sulthan Bathery, Kalpetta, Mattancherry, Ernakulam, and Kochi of Kerala and Gudaloor of Tamil Nadu.

The data collected was used in developing the cost sheets giving a full-length view of the costs incurred at every activity by the bearer. The cost sheet makes us understand at which node what amount of capital is involved to optimize the procurement for a new enterprise. The questionnaires were also filled personally by meeting the suppliers' in-person.

4.3.4 Data Analysis

Percentage and tabular representation were used as statistical techniques for data analysis. The percentage costs incurred out of total operational losses are analyzed by using simple percentages. The supply chain models and procurement models are also made with the help of the data collected.

4.4 Results and Discussions

This section explains the procedure for the supply chain of black pepper along with its cost at each node of the supply chain for Mattancherry Market and SulthanBathery, Kalpetta, and Gudaloor Market. In many cases, the masala companies tend to procure the collector quality pepper to process it in their manufacturing facility and make use of the by-products also. The suppliers collect the berries from the collectors and pay the required commission and margin to the collectors. The manufacturers or suppliers process the lot as per the customer's specifications.

4.4.1 Mattancherry Market

The lot undergoes a series of processes required for making different grades of pepper. Each process in the series carries a cost with it. The processes followed for making different grades of the pepper berries are as follows:

- Varal—It means to lift with hands.
- Kuttu—It means to hit with any instrument.
- Navari—It means the lot name.

The nodes involved in the supply chain are as follows:

- *Farmers/cultivators/producers*
- *Brokers/agents*
- *Collectors*—who procure at a local level in small quantities and supply to wholesale traders
- *Manufacturers/Suppliers*—who buy stock from the collectors and in some cases directly from the farmers and sell to masala companies after transforming the stocks as per their customer's specifications.

After receiving the order from the collectors, the bags are unloaded and stacked. The labor charges vary from market to market. In Kerala, labor charges are based on a piece-work system wherein the charges are paid by the number of piece-works performed by the worker and not by a fixed payment. The market has its labor union. The pepper trade is regulated by an association named Indian Pepper and Spice Trade

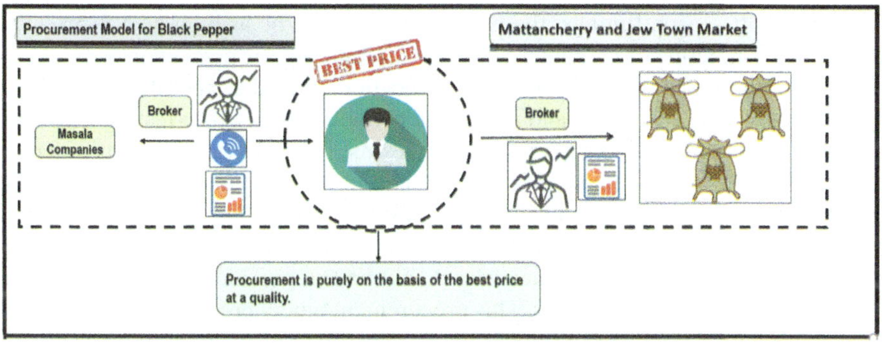

Fig. 4.1 Procurement model for black pepper (Mattancherry Market). *Source* Author

Association (IPSTA). The entity is composed of the giant suppliers of the market. The Mattancherry market works on a broker model. Wherein every activity apart from processing, ensuring standards, and dispatching the order to the customers is done by the brokers. The brokers make sure that the desired quality, quantity, time, credit period, etc. are maintained, and the delivery is done at the required time for which they charge a percentage commission out of the total trade. This adds another additional screen to the supply chain. The procure model for black pepper of Mattancherry Market is explained in Fig. 4.1.

The manufacturers perform the desired processing required to generate the customer grades. The processing converts the collector's quality or ungarbled pepper to garbled pepper. The ungarbled pepper has an inferior quality as it is less clean. The dried berries may contain a fungus named *phytophthora* usually during rainy seasons and require water-washing leading to an increase in operational loss percentages. The water-washed berries are sun-dried or dried in mechanical driers based on the lead time and the dispatch date. The dried berries are graded by the traders into different grades based on the bulk density, black color, pungency, moisture content, dia. of a berry, volatile oil content, piperine content, etc.

The most common grades for Indian pepper available in the Mattancherry market are

- TGSEB—*Tellicherry Garbled Super Extra Bold*
- TGEB—*Tellicherry Garbled Extra Bold*
- MG1—*Export Grade.*

After the grading is done, the by-products are also separated as they fetch a good economic value, for example, the Black pepper husk or spent, light berries, pest-infected berries, pin-heads, etc. The by-products are exported to many countries and oleoresins manufacturers. The percentage loss during the entire processing cycle starting from water-washing the dried berries to packaging in a specified packaging system by the customer depends on the weather conditions, supply, maturity of the berries, crop status, and many more factors. During the seasonal times in the market, the percentage loss in the entire processing is 2.5–3.0% of the total lot size processed.

During the off-season timings, the percentage loss can be as high as 4.5–5%. The loss incurred in terms of volume is added by multiplying the percentage loss factors with the value of per kg pepper. The certification from the laboratories is also a requirement of the customers and is fulfilled by the suppliers. The testing is usually done by an outsourced laboratory that certifies the lot with specifications given by the suppliers. The specifications are mentioned in the agreements of the purchase made between the suppliers and any other B2B entity, usually a masala company.

The market involves brokers at the two nodes of the supply chain. This procurement model helps those companies who do not have a good idea about the market, have communication barriers, etc. This procurement model provides the customer to order by staying at their location, the broker takes care of the quality, lot size, and delivery timings. The negotiation done is quite satisfying but in certain cases, the brokers may not sell some material at a time to purchase for earning a better deal from somewhere else, the masala companies face issues. The companies need to completely rely on the brokers. The flow diagrams of the black pepper supply chain (owned laboratory facilities and outsourced laboratory facilities) for this market are depicted in Figs. 4.2 and 4.3, respectively.

The value addition along the supply chain decreases as we go from the farmer's end toward the supplier's end. The value addition again increases and becomes the highest in the retail sector wherein the product is made available near the residential areas. The supply chain has a higher total cost than the one without brokers.

Most of the procurements by the masala companies happen to be from January to April. During these four months, the Mattancherry market supplies black pepper to almost the whole world. The supplies from other Indian states Karnataka, Tamil Nadu, and other countries like Vietnam, Sri Lanka, Indonesia, Brazil, etc. hit the

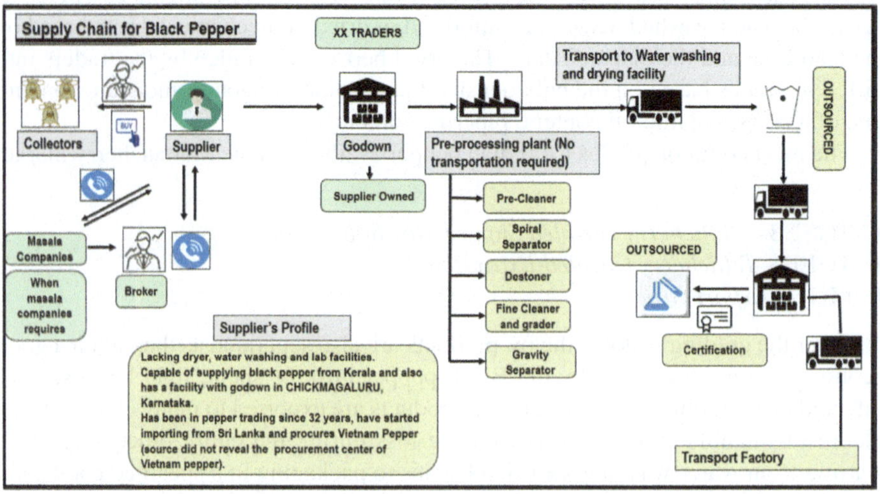

Fig. 4.2 Supply chain models for black pepper (owned laboratory facilities), Mattancherry Market. *Source* Author

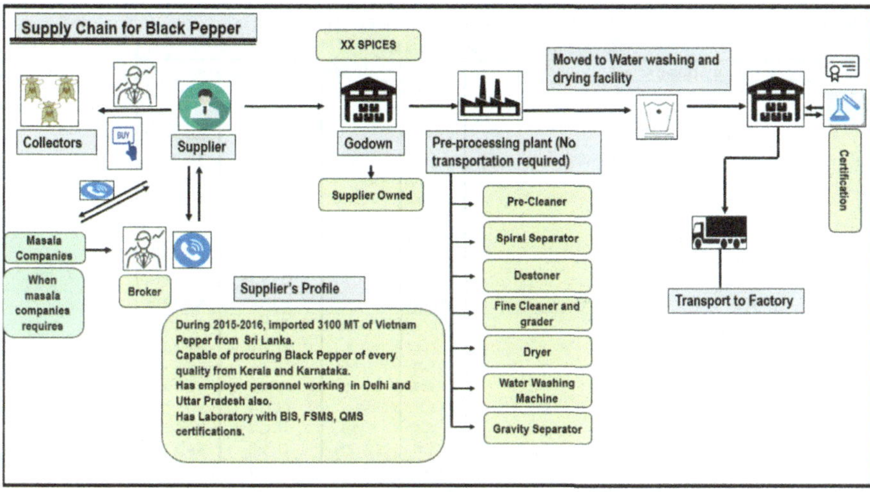

Fig. 4.3 Supply chain models for black pepper (outsourced laboratory facilities), Mattancherry Market. *Source* Author

market post-April of year. The purchases during the rainy season are often ignored because of unfair prices, disturbances in distribution channels, and hoarding. The crop has a very volatile market and has shown many up and downtrends over a decade irrespective of the government interventions. It is ideal to purchase during the starting 4–5 months of the year. The credit period in this model was 30–45 days after the purchase date. The lot dispatches once the customer has been sent a mail containing the soft copy of a laboratory certificate. The cost sheet for the procurement model is made using MS Excel. The cost sheet explains each cost incurred at every node of the supply chain which is exhibited in Table 4.1.

The cost of procurement is less than the Mattancherry market as the region has low labor charges, less number of buyers, and uncertainty of the crop. The buyers and sellers are lesser in number than the Mattancherry market. Unlike the Mattancherry market, this market also possesses traders who just sell the garbled uncleaned black pepper at a very low rate. Some masala companies like to procure this kind of pepper to process it in their facilities and to make use of the by-products also.

4.4.2 *Sulthan Bathery, Kalpetta, and Gudaloor Market*

The market established in Kalpetta and Sultan Bathery, Kerala, offers a mixed procurement model, wherein a model without the broker's intervention also exists. The market is flooded with suppliers and small collectors and offers good quality produce all over the year. The labor charges in these cities are different from the Mattancherry market and influence the final cost as well. The flow diagram of the

Table 4.1 Cost sheet for black pepper whole (Mattancherry Market)

Black Pepper - May 2018					
Purchase Cost		Masala Companies			
Bag size - RM		1 bag		50.00	
Bag size - FG		1 bag		50.00	
BASE RATE	rs/kg			380.00	
Varal, Filling And Weighing	1 Qtl	7.45	0.07	380.07	
Double Lined Gunny Bag	rs/bag	105.00	2.10	382.17	
Double Stitching	1 bag	2.00	0.04	382.21	
Single Side Marking	1 bag	1.15	0.02	382.24	
Loading	1 bag	4.6	0.09	382.33	
Gadi Bhada & Pala	rs/kg		2.50	384.83	
Collector's Margin	rs/kg		10.00	394.83	
RM_COST_SUPPLIER	rs/kg		14.83	394.83	
Unloading	1 bag	3.45	0.07	394.90	
Atti	1 bag	2.3	0.05	394.94	
Bulking	1 Qtl	5.15	0.05	395.00	
Rassi	1 Qtl	3.45	0.03	395.03	
Varal with filling	1 Qtl	4	0.04	395.07	
Weighing	1 Qtl	3.45	0.03	395.11	
Local Delivery	1 Qtl	17.25	0.17	395.28	
Supplier's Labour Charges	rs/kg		0.45	395.28	
Machine Cleaning	rs/kg		2.00	397.28	
Labour for by-Products Realized	rs/kg	0.50%	0.22	397.50	
Filling & Weighment	1 Qtl	7.5	0.08	397.58	
Double Stitching & Single Side Marking	1 bag	3.15	0.06	397.80	
Loading	1 bag	4.6	0.09	397.89	
Freight to Drier Facility	rs/kg		0.50	398.39	
Drying Charges	rs/kg		1.35	399.74	
Unloading	1 bag	3.45	0.07	399.81	
Operational Losses (in-season)	%	2.50%	0.45	400.26	NOT USED
Operational Losses (off-season)	%	4.50%	0.80	400.61	
Lab_Testing	rs/kg	30	0.60	401.21	
Loading	1 bag	4.6	0.09	401.30	
Freight To Delhi	rs/kg		7.50	408.80	
Supplier's Total Cost	rs/kg		28.64	408.80	
Supplier's Margin	%	3%	12.26	421.07	
Brokerage	%	1.50%	6.32	427.38	
			Total	427.38	

Source Author

black pepper procurement model and supply chain for these markets are depicted in Figs. 4.4 and 4.5 respectively. The supplies to this market come from Wayanad and Idduki, Kerala, and Kodagu in Karnataka. Most of the suppliers in these markets own their plantations of pepper, clove, and cardamom. Gudaloor in Tamil Nadu is also a big production center of black pepper. The market has a decent number of suppliers but the capacity levels that factory owners require are a bit less with the suppliers. The cost sheet of the black pepper value chain is exhibited in Table 4.2.

The black pepper markets in these areas are a bit cheaper than the Kochi market, as all the varieties of black pepper are not available like the Kochi market. The brokers engage at only one node of the supply chain, unlike the broker model.

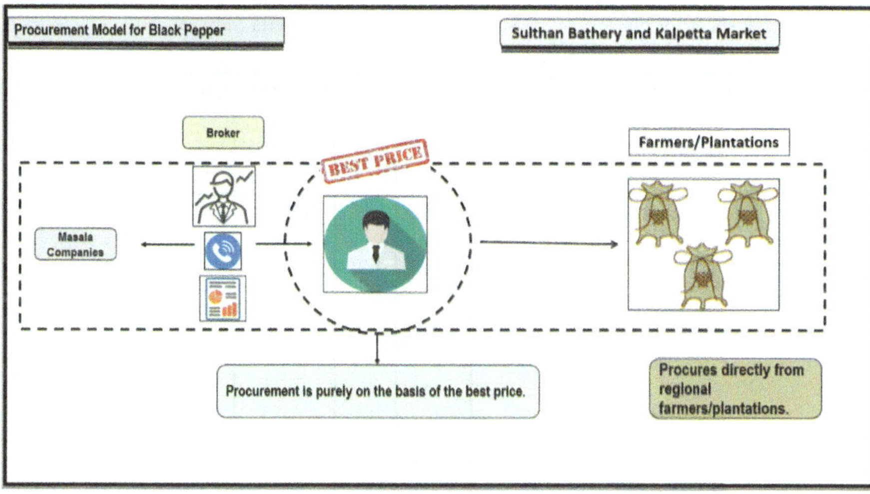

Fig. 4.4 Procurement model for black pepper (Sulthan Bathery and Kalpetta Market). *Source* Author

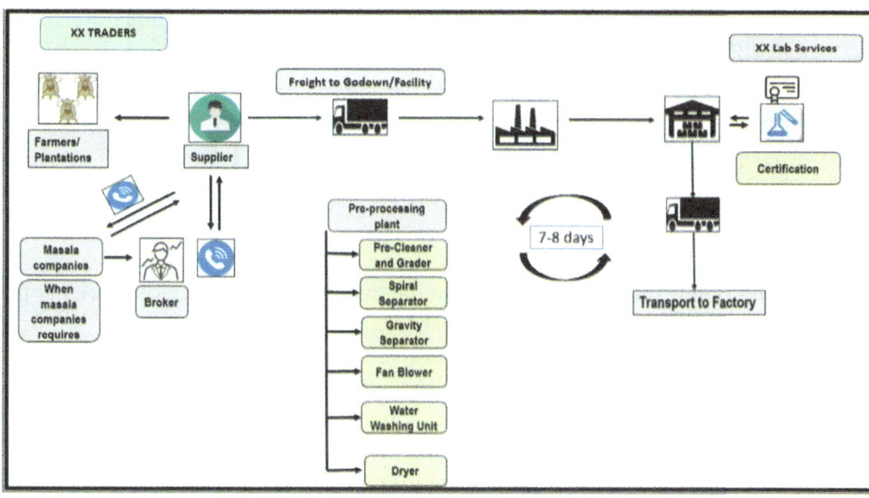

Fig. 4.5 Supply chain model of black pepper (Sulthan Bathery and Kalpetta Market). *Source* Author

4.5 Conclusion

The present work provides aid to the new enterprises in devising their strategies, procurement model, and supply chain for black pepper procurement. Black pepper is a very volatile commodity and highly dependent upon weather conditions. Sometimes, due to extreme weather conditions, black pepper production gets affected, and this

Table 4.2 Cost sheet for black pepper (Sulthan Bathery and Kalpetta Market)

Black Pepper May 2018				
Purchase Cost		**Masala Companies**		
Bag size - RM		1 bag		50.00
Bag size - FG		1 bag		50.00
BASE RATE	rs/kg			380.00
Freight to godown	rs/kg		2.50	382.50
Unloading	1 bag	8.50	0.17	382.67
Stacking	1 bag	3.30	0.07	382.74
Weighment and Pouring	1 bag	3.8	0.08	382.81
Machine Grading	rs/kg		2.90	385.71
Drying	rs/kg		1.20	386.91
Operational Losses	%	2%	3.28	388.99
Weighment and Filling	1 bag	10	0.20	389.19
Double Lined Gunny Bag	1 bag	105	2.10	391.29
Double Stitching	1 bag	1.8	0.04	391.32
Nylon Thread	1 bag	0.25	0.01	391.33
Lab Testing	rs/kg		0.60	391.93
Loading	1 bag	8.50	0.17	392.10
Freight To Delhi	rs/kg		6.20	397.53
Supplier's Total Cost	**rs/kg**			**397.53**
Broker's Margin	**%**	**1%**	**0.10**	**397.63**
Supplier's Margin	**rs/kg**		**25.00**	**422.63**
			Total	**422.63**
Total labour cost	**rs/kg**		**0.72**	
Total Transportation Cost	**rs/kg**		**8.70**	

Source Author

drop in production results in the increase in the average annual price of black pepper which can be seen in prices of a few years back.

This increase in the black pepper prices influenced the Indian manufacturers, traders, and wholesalers to import more and more black pepper from around the world, increasing the annual imports. This shows that black pepper procurement is a tough decision when it comes to new entrants who want to enter the market. The new enterprises may face problems in such situations as they would not be having enough penetration in each market for pepper. The models discussed in the chapter namely the supply chain models of Mattancherry, SulthanBathery, and Kalpetta show that procurement from the Mattancherry market was a bit costlier than any other procurement centers as the market operates with the involvement of brokers but provides a better quality and food safety options than other markets. The cost sheets will help an entrepreneur in coming to a fair deal with the traders during the procurements. When it comes to choosing a procurement market, the proprietor should understand its capability and standards to satisfy the goals of the enterprise. Every market has its pros and cons. The procurement decision becomes very dynamic when it comes to procuring black pepper which is an agricultural commodity dependent on numerous factors.

The market is the focus of all activities in a given chain. It is important to recognize that all stakeholders involved in a supply chain must work together and align their activities to meet the needs of the end customer.

4.6 Limitations and Future Scope of Work

The procurement market for black pepper is unorganized, and due to this, data regarding the processes varies among traders. Further, the study is limited to few markets, whereas too many small markets are also there. The trader played a bigger role in the black pepper supply chain but these traders are not concentrated around a region but are scattered across India. The sample units used in the report are chosen based on the arrivals, sales, demand, technological capabilities, capacity, and ability to supply supplies to factories.

The study can be further extended for other high values spices like green cardamom and clove, saffron, cinnamon, etc. The study can be done from the producer's point of view regarding the investigation of increasing producer share as the black pepper process requires lots of time and effort to reach the market.

References

1. Thangaselvabal T, Gailce Leo Justin C, Leelamathi M (2008) Black pepper (*Piper Nigrum* L.) 'The king of spices'—a review. Agric Rev 29(2):89–98
2. Krishnakumar PK (2019, October 29) Black pepper prices lowest in a decade. The Economic Times. https://economictimes.indiatimes.com/markets/commodities/news/black-pepper-prices-lowest-in-a-decade/articleshow/71799047.cms
3. Food value chain partnerships—end to end approach (2020). ASSOCHAM. https://www.assocham.org/userfiles/Food%20Value%20Chain%20Final-min.pdf
4. Sharangi AB, Acharaya SK (2007) Performing supply chain in banana, black pepper, capsicum and seed spices: status, issues and scopes. J Crop Weed 3(2):47–51
5. Hema M, Kumar R, Singh NP (2007) Volatile price and declining profitability of black pepper in India: disquieting future. Agric Econ Res Rev 20:61–76
6. Kuruvila A, Sabu SS, Manojkumar K (2019) Price behaviour of black pepper in Indian and international markets: a comparative analysis. J Spices Aromat Crops 28(1):27–33. http://doi.org/10.25081/josac.2019.v28.i1.5741
7. Paulrajan R (2012) Spice route: logistic journey of spices in retail supply chain perspective. UTMS J Econ 3(1):9–20
8. Prabhavathi Y, Krishna Kishore NT, Seema (2013) Analysis of supply chain of spices in India: a case study of red chilies, vol 2, no 9. http://www.ijsrp.org/research-paper-0913/ijsrp-p21101.pdf
9. M S Y, Mokshapathy S (2013) Production and export performance of black pepper. Int J Humanit Soc Sci Invention 2:36–44
10. Ganpathy MS, Naveenkumar HS, Murthy MM, Rajshekaramurthy S (2014) Value chain analysis of black pepper in Karnataka. Int J Sales Mark Manag (IJSMM) 3(2):13–18

11. Dessie M, Woldeamanuel T, Mekonnen G (2017) Value chain analysis of red pepper: the case of Abeshge District, Guragie Zone, South Ethiopia. Int J Environ Sci Nat Resour 2(3). http://doi.org/10.19080/IJESNR.2017.02.555590
12. Majo G, Elsa C (2017) Emergent global marketing challenges for Kerala cardamom producers Vis-à-Vis the role of the spices board of India. Int J Community Dev Manag Stud 1:39–62
13. Cristino Mandinga Bonfim F, José Castro C, Fernando Brito S, Augusto Manuel Nogueira Gomes C, Zélia Maria Gonçalves S (2020) The organic pepper (*Piper nigrum* L.) value chain in São Tomé e Príncipe under a value chain analysis for development methodology perspective. Direct Res J Agric Food Sci 8(4). http://doi.org/10.26765/DRJAFS40252693
14. Krishnaraj S, Gunaseelan R, Arunmozhi M, Sumandiran CSP (2020) Supply chain perspective and logistics of spices in Indian retail industry. Mater Today Proc S2214785320314474. https://doi.org/10.1016/j.matpr.2020.02.681
15. Friesenbichler KS, Kuegler A, Reinstaller A (2021) Does value chain integration dampen producer price developments? Evidence from the European Union. World Econ 44(1):89–106. https://doi.org/10.1111/twec.12993

Chapter 5
Sustainability in Food Supply Chains: Case of Southeast Asia

Modisa and Thappasarn Jaikaew

Abstract Food products go through a chain of various markets and companies to be approachable for human consumption; this phenomenon is termed as food supply chain. With the demands of information about the food from farm till it gets to consumers, food supply chains worldwide have opted newer ways to function, allowing consumers to control what they eat. Most of the Southeast Asian countries are making their way toward getting into the race of developed economies, where food supply chains contribute to the countries' economies by the export business. With the food chains, food safety is one of the major concerns that can bring damage to whole export systems; therefore, various technologies and better practices are introduced. In this chapter, a study was conducted at a coffee and bakery shop in Bangkok, Thailand, to understand the supply chain and various food safety measures being taken at its scale. Moreover, every resource has limits, and if exploited, it can result in scarcity of the same resource. Careless production and processing practices of food products can pose threats to the sustainability of various other natural resources; food waste is one of such major threats; therefore, various possible ways of food waste management from various literature are enlisted in this chapter.

Keywords Sustainability · Food supply chains · Coffee and bakery · Food waste management

---

Modisa · T. Jaikaew (✉)
School of Environment, Resources and Development, Asian Institute of Technology, Bangkok, Thailand
e-mail: thappasarn.j@mail.rmutk.ac.th

T. Jaikaew
Department of Food Technology and Nutrition, Faculty of Home Economics Technology, Rajamangala University of Technology Krungthep, Bangkok, Thailand

R. S Mor et al. (eds.), *Operations and Supply Chain Management in the Food Industry*, Lecture Notes in Management and Industrial Engineering,
https://doi.org/10.1007/978-981-16-5555-5_5

5.1 Introduction

The supply chain (SC) is the process used by companies for creating and distributing its products or services to the end consumer. Supply chain action begins at the point of demand creation and works toward fulfilling that demand with efficiency and efficacy [22]. An efficient supply chain can help a company avoid expensive recalls and customer complaints [12]. In addition, Lu [14] has defined SCM as simply the business management with a relevant and specific supply chain viewpoint. A food supply chain typical consists of players like producers, processors, distributors, retailers, etc. Retailers help the producers for their promotional activities, too [2, 7].

Nowadays, many countries in the Southeast Asia region are interested in food supply chain because food and agribusiness are of increasing importance to the region. The region's diverse markets include some of the world's largest agricultural exporters such as Thailand and Vietnam, as well as innovative centers of research and development such as Singapore. Although rice is still the region's main crop, other commodities such as maize, coffee, cocoa, and fruits and vegetables are also important. Some member states are also specialized in fresh and canned fish or in livestock. Besides, palm oil is one of the main agricultural products for both Malaysia and Indonesia [15].

Consequently, this chapter aims to review the state of food supply chains in Southeast Asia with a focus toward sustainability and to present case study of food supply chain of a coffee and bakery shop in Bangkok, Thailand.

5.2 Literature Review

For southeastern Asia, food and agribusiness have greater value as it contributes to the large proportion of the economy in each region. For example, Thailand being the biggest exporter of rice, shrimp, and rubber, Malaysia ranks first in palm oil export in the world. In contrast, Indonesia occupies the second position in palm oil and rubber export, Malaysia is also third largest rubber exporter, and fifth for the cocoa, with Indonesia on third position for cocoa production; moreover, Philippines has the highest exports of sugar and pineapple. Vietnam has the second largest exports of rice and coffee in the world. But there are some issues highlighted by FAO for the exports of Southeast Asia countries, such as 20–30% loss of harvest during transportation, demands from developed economies for quality check of the products, and lesser control of governments over regional logistics and infrastructures for the sake of improvement [15]. In order to know more about the food supply chains of Association of Southeast Asian Nations (ASEAN), following are some of the areas highlighted for food supply chain management.

5.2.1 Partnerships

The food supply chains of SEA region are continuously changing, which includes strong relationships between all the members within a supply chain and with members of other food supply chains. These relations can be partnerships and alliances with competition replaced by joint-projects and deals. Partnerships, instead of having solitary supply chains, strengthen the resources of each participating organization. On the other hand, consumers are benefitted by reduced costs, enhanced availability of products of desire and better quality. For example, a local dairy firm of Thailand cooperated with Danone (French multinational food product company) and both of them are in symbiotic relationship, getting benefits from each other's strengths and their cooperation is experiencing continuous rise [15].

Another successful example of modernization of food systems in SEA is Thailand's top supermarket chain which understood the global movement of fresh produces. Businesses (Tops, Ahold Thailand), research institutes such as Kasetsart University, the Netherlands, and government and semi-governmental corporations worked together in this project to develop better fresh products' chain by using "preferred supplier approach" to reduce the number of suppliers and using good agricultural practices (GAP) and GMP throughout the supply chain. Moreover, lead time was reduced significantly, which allowed for greater accuracy and JIT (just in time) to reach to 98% [3].

On the other hand, Singapore, being a small country with limited resources, imports 90% of the food items, and the import is just from one or two countries; hence, if there is any disturbance in the trade, then it will have adverse effects on Singapore. For this reason, Singapore has made decision to concentrate her R and D departments on finding ways to increase the production of fish, eggs, and leafy vegetables. The International Enterprise (IE) and Agri-Food and Veterinary Authority (AVA) of Singapore are working mutually to find new ways of production and find new agribusiness opportunities with other nations worldwide. As stockpiling of important food products is helpful and easily implantable in local markets, Singapore is looking at substituting the important dietary items with their liquid or powder forms, such as powdered or liquid eggs, frozen pork cuts in place of fresh chilled pork meat [24].

Although some SEA countries are opting to adopt new technologies and introduce modern ways into the food supply chain, some others still face the issues of delays by the administrating laws and regulations, which in turn affect the costs of circle of trade and logistics. One such example is the Indonesian food supply chains. As per World Bank report, the processing of some Indonesian products, such as shrimp and pineapple, is done abroad while 10% of exports are sent with greater lead time and cannot make up to the regional shipment ports on time. As the main mean of transport is trucks which spend 75% time at customs, warehouses, lift queues and logistics and trade are still paper-based [5].

5.2.2 Technology

The infrastructures of a country have huge influence on supply chain management. Use of newer technology is on trend in the supply chains of Southeast Asia (SEA). Modern tools such as, electronic, resource planning (ERP) software in place of MS Excel spreadsheets, are being widely used in food business to update the inventories and data from suppliers and customers records. Secondly, newer technology of radio frequency identification (FRID) is finding its way in SEA countries' food chains for quicker and more accurate tracing of the product from the product arriving at the plant till it reaches the international consumers in Japan EU and USA. This technology uses automated or electronic tracking devices to decrease the costs associated with labor used for paper-based tracking. And if recall is needed, RFID will provide aid through saved information of origin of the food, processing, and expiry data. One such system was used in a study in Korea for the traceability of beef, starting from the live animal's identification until the beef steaks' packaging was obtained by slaughtering the animal and further processing the meat. The systems gave a serial number called RF chip to each package with the animal ID [23]. Moreover, the green productivity philosophy is the innovative idea implemented to reduce transaction costs, wastes throughout the supply chain, and delayed deliveries [15]. Like Malaysia, traditionally an agricultural country but is converted into an industrial country, it has focused on trade and foreign investment for years. The logistic sector is given much importance in the ninth Malaysian plan and third Industrial master plan (IMPT3), to achieve competitiveness globally. Moreover, in 2009, electricity and telecommunication was major problem of all the supply chains of Malaysia which was overcome by 2012 [5].

Thailand is an important exporter of frozen shrimp products worldwide, which earns her a great amount of revenue (with 305,807 tons of export in 76,036 million baht in 2006), is trying to embrace modern techniques to gain the trust of large international markets. And with the funding from National Innovation Agency, Thai exporter uses RFID for better traceability which dominates Thai seafood exports over their competitors. A study on the strategic development of supply chain management of Thailand showed shrimp stakeholders applying modern management techniques in various sector of their businesses such as customer relationship management, order-cycle time management, product flow, and development of products [15].

5.2.3 Food Safety

According to Othman [16], the Vice-Chairperson of the Codex Alimentarius Commission, the production, processing, and marketing systems of SEA countries comprise small to large scale in range the product goes through several steps and handlers in the chain. There is a lack of knowledge and expertise about modern ways and technology usage (in some of the ASEAN member countries) and lack adequate infrastructure. The major source of food in this region is the street food vendors, which is accessible and cheaper, but this sector hires unskilled or semi-skilled employees posing a risk to the safety, hygiene, and sanitation of food processes and product. Moreover, there is lack of concern about national food safety problems and knowledge about the effects of contamination of food on health and the economy. There is lack of cooperation among agencies for food safety due to political reasons. Therefore, food safety policies are either not existing or low primacy in some of the countries situated in the SEA region. Furthermore, the constituents and main concerns of food safety control systems vary among the countries, including food control management, services for food inspection, legislation for food safety, laboratory testing services and information, training, education and communication.

The Association of Southeast Asian Nations (ASEAN) is taking initiatives to address the food safety issues in supply chains of food items sold locally or exported. Various groups are created in this regard, such as ASEAN Expert Group on Food Safety (AEGFS), which works on facilitating and coordinating food safety in ASEAN, and ASEAN Food Safety Improvement Plan (AFSIP), which functions in the implementation and improvement of food safety policy and plan of actions. Of above-mentioned food safety concerns, five areas are considered to be most important for food safety, which are assigned to five ASEAN countries to work on for implication and further betterment, which includes Malaysia, for the inspection, monitoring, and surveillance of food and certification; laboratory testing for Singapore. Consumer participation and empowerment are assigned to Indonesia, Thailand, to share information in the food network of ASEAN and, finally, Philippines, for the legislation.

Secondly, the cooperation program between EU and ASEAN deals with food quality, standards, and conformational assessment of the products exported internationally from all the member countries of ASEAN, which also works on the major areas of food safety along with implication of HACCP, GMP, GHP in Small and Medium Enterprises (SMEs). This cooperation has led to the development of laboratory networks such as microbiology in Vietnam, pesticide residues and mycotoxin testing in Singapore, heavy metals and trace elements in Thailand, and genetically modified organisms (GMOs) in Malaysia [16].

5.2.4 Green Supply Chain Management

Green supply chain managements are trending in the food industries of SEA region, through which various companies show their sincerity toward sustainability [15]. Green supply chain initiatives taken by food industries of SEA countries include reduction of waste, substitution of material by environmental friendly raw material, recycling, low density packaging, and biodegradable packing. It also includes reduction of operating cost and better services for customers [19].

5.2.5 Packaging and Transportation

In ASEAN area, the food products are packaged in boxes with grooves and ridges, and plastic crates that can be returned or reused, replacing the traditional packages such as baskets and wooden crates. The plastic packages are used for local high-end markets and institutions while ridged and corrugate boxes are used for food products going for exportation. Plastic bags recently replace these packages once they reach wholesale markets because plastic bags are cheaper, easy to handle, and available. But these bags cause environmental sustainability problems because these are not biodegradable and decrease the shelf life of products due to insufficient availability of air and provide lesser protection against mechanical damages. Moreover, the retail packaging has grown due to consumer demands for quality and supermarkets, where fresh produce is also sold with branding in high-end supermarkets.

For example, Philippines has the highest use of plastic bags. Fresh produce is not given much priority, and to reduce the expenses of transportation, the fresh vegetables and fruits are transported in large containers which can cause damage to the produce due to compression. Like, papaya is shipped in large wooden crates from Mindanao and eggplants are shipped in large but overfilled kaing/bamboo baskets and woven polypropylene (WPP) sacks from Iloilo which is a province in Visayas [6].

Whereas, in Thailand, the fresh vegetables and fruits are distributed by loaded pickup vans and trucks directly without any packaging and at wholesale market, purchasers receive their orders in bulk containers like plastic crates, bamboo baskets, or plastic bags. The damaged commodities are sold at low prices. Some produce can also be packaged in plastic bags or containers manually before loading into the truck; this depends on the distance to the destination. Moreover, for bulk packaging, mostly trapezoid plastic crates are used which contain handles made of steel. The trapezoid shape helps in storing the empty crates in nested piles. The exporters and large retailers also use round and rectangular plastic crates for fresh vegetables and fruits which can be nested. Apart from these, some other types of packaging material, such as net bags, paper bags, cartons, wooden crates, trays, and film wraps, are used depending on the type of produce.

In case of Vietnam, similar types of packaging material are used depending on the type of produce but the transportation means may vary depending on the distance from the delivery point. Like, transport of products from farm to farmhouse or shed is done by hauling carts, single-wheeled carts, and bicycles. In contrast, delivery to a farther destination is carried out either through motorbike, bicycle or three-wheeled carts. Traders make use of motorbikes, pickups, boats, and trucks to receive their orders from suppliers. Vietnam has set regulations for the exporting products, including suitable packaging in particular type and size of packaging material, appropriately labeled with complete instructions written in two languages of English and French, along with name of product, quantity, and address of manufacturer country of origin. Additionally, if the package has direct contact with the food product, only certified wrapping needs to be used. The packages are examined by Vietnam Food Administration [21].

5.2.6 Urbanization

Furthermore, in Asian-Pacific areas, people migrate to urban regions for better lifestyles, which cause rapid increase in the spread of supermarkets. And in the next twenty years, these regions' urban population will increase by half a billion, which is more than half of the total area [15]. Although the channels of food marketing in SEA are changing from shops to supermarkets and hypermarkets, still a great deal of fresh food is sold as wholesale markets, which is speedily evolving in Bangkok through the government's policies for food supply chains and interactions between shareholders [15].

Another study shows that supermarkets are expanding rapidly in whole Asia, starting with more developed countries such as Taiwan, Korea, Hong Kong, and Japan. The supermarket chain includes both the international names such as Walmart, Big C, Tesco and those developed locally. Some local supermarkets also expended to other Asian countries such as the chains from Hong Kong and Japan expended to Malaysia, Indonesia, and even the supermarkets from Vietnam spread to Cambodia. Although such international chains of supermarkets have not targeted Myanmar but the local supermarkets developed in the country are relatively advanced, such as City Mart, K-Mart, Super-One. These supermarkets contain the local goods along with imports from ASEAN, China, and India [25].

5.2.7 Diet Transformation

It has been observed that the diet has been transforming gradually in whole Asia including countries in the Southeast Asian region. The main reasons for this transformation are income-induced diet modification. Higher-income people tend to pay for more convenient food and include a variety of food items rather than on a typical list of affordable food products. Secondly, adoption of more western style diet as consumers have shown more preference for McDonald's, Starbucks, etc. along with protein-rich foods such as meat and fish, temperate zone food items such as apples and processed products and drinks. Traditionally rice preferring societies are consuming wheat products such as bread, pizza, pastry, and cakes, due to which more and more supply chains are adapting to changing diet styles and focusing on such products to be readily available for the customer. A report showed that major rice-consuming countries of SEA have seen a decline in rice consumption, like between 1981 and 2001, Thai diet saw fall in rice consumption from 60 to 43%, while wheat consumption tripled [11]. Moreover, the consumption of potato has also increased ten times since 1981. Due to such dietary evolutions, the supply chain systems are also changing, and a study of FAO on agriculture from 2015 toward 2030 shows that SEA countries have shifted from net exporters to net importers of food products which may grow in the future [18].

5.2.8 Effect on Food Supply Chains Due to Natural Disasters

Sustainable environment has been seen to have greater impact on supply chains. And the unstable environmental conditions are one of the major issues hindering the success of food supply chains in SEA countries [5].

According to a study, the unsustainable environmental conditions have caused great losses to all types of businesses. For example, the infrastructure in Indonesia is affected by flood from time to time. And according to the chairman of Indonesian Employers Association, Sofjan Wanandi, the businesses on land of Indonesia are facing losses because of disturbed supply chains which in turn cause reduced production and 300 factories situated in East Jakarta had to stop their productions because of interrupted power supply and higher costs [5]. Secondly, in 2011, World Bank mentioned in their report that due to blocked sewer and drainage systems because of garbage, water stayed at facilities for months which caused many large-scale manufacturers to stop their production, and most of the uninsured production systems showed adverse effects on the food supply chains globally [17]. On the other hand, due to floods, the two main problems of Thailand's food supply chain are energy expenses and human labor cost [5].

5.3 Case Study of Food Supply Chain

To understand the dynamics of food supply chain, a case study was carried out for a coffee and bakery at Bangkok, Thailand. The same is discussed further in this section ahead.

5.3.1 About "Coffee Moona and Bakery"

The information of Coffee Moona and bakery is described below. First of all, it will come up with general information of the business;

The Coffee Moona and bakery was established by Ms. Sasikarn Thuenprom in 2013 at her house, situated at 244 Moo 3 Krathum Rai, Nong Chok, Bangkok, postcode 10530, Thailand. We came to know about Ms. Thuenprom and her shop from her daughter, Miss. Nichakarn Thuenprom, who studies at Asian Institute of Technology.

Ms. Thuenprom received her bachelor degree in Business Administration (B.B.A) with specialization in Marketing. After her graduation, she started working at the marketing department of Adidas brand. Later, she got interested in learning baking and received certificate of baking from one private school and another government school (The Vocational Training Center, Nong Chok, Bangkok) and opened her own coffee and bakery shop by using her savings. She did not get any kind of support from the government.

One of the reasons to open the shop was to be independent and work as she likes to; secondly, she wanted to excel over her interest in baking and provide newer and innovative product for the customers. Therefore, she selects all raw materials for the products by herself.

Presently, the products sold at the shop are beverages (such as coffee, chocolate drinks, Thai milk tea, and Italian soda), bakery products (such as egg tart, butter cake, and brownie, Chinese spring rolls), and also bakes cakes based on pre-orders for various occasions, such as birthdays, New Year, and valentines. The price for beverages ranges from 25 to 30 ฿ (TBH) and 10 to 30 ฿ for bakery products.

5.3.2 Supply Chain of Bakery

Food supply chain for Coffee Moona is shown in Fig. 5.1. Supply chain map shows the starting point of suppliers from whom the raw material is bought till the final products presented to the customers. Overall, there are four main suppliers of raw material, among which, the coffee beans are ordered from Nan Duo Coffee Limited Partnership under brand Phu coffee at Nan province. The Phu coffee is a local Thai brand located at the Northern part of Thailand. The brand owner also owns coffee farms for the coffee beans to be cultivated for the brand, which produce enough beans

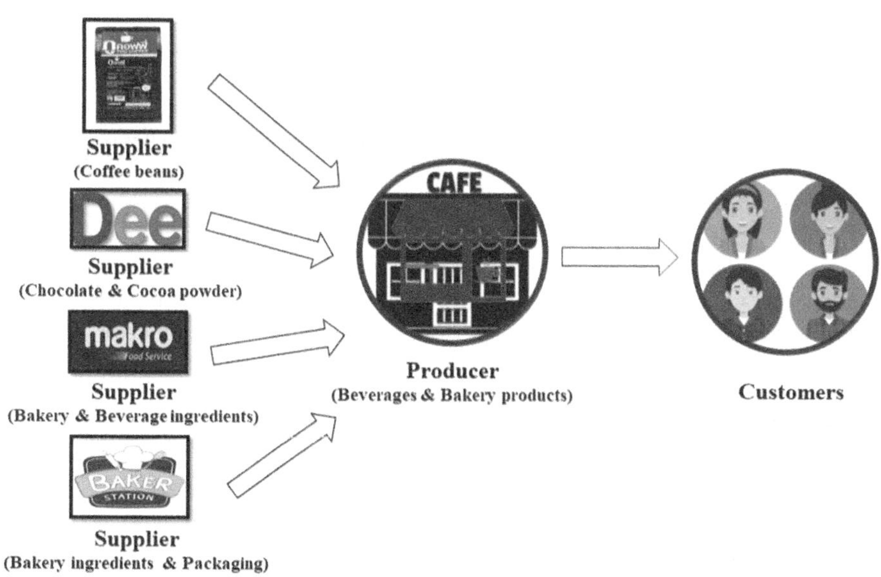

Fig. 5.1 Map of food supply chain for Coffee Moona. *Source* Authors

for the product but if the cultivar is not sufficient then the beans are obtained from trusted contract based coffee farms in the same province. The brand's coffee beans processing plants are certified for food safety and quality control from Thai-FDA approved Notification of the Ministry of Public Health (No. 197) B.E. 2543 (2000) Re: Coffee and also Halal food certified by the Central Islamic Council of Thailand (CICOT). Secondly, chocolate and cocoa powder are bought from an online website called Dee.com which sells products certified by international food standards such as GMP and HACCP. Furthermore, some basic ingredients for bakery and beverages (like concentrated fruit juice, soda, eggs, sugar, milk, etc.) are shopped from Makro food services, a wholesaler registered for its food safety concerns at Thai-FDA. The selected raw materials bought from Makro foodservice are also certified by GMP, HACCP and Halal food. Lastly, a shop situated across Coffee Moona under the name of "Baker station" supplies some basic bakery ingredients, such as flavoring agents, food colors, flour, and butter, along with packaging services for the products of the bakery. Goods from the Baker station are certified by GMP and HACCP under Thai-FDA.

By using all these ingredients, various bakery products and beverages are prepared at the shop and sold directly to consumers. The products are prepared from ingredients which are safe for consumption, and Ms. Thuenprom showed greater concern about customers' health.

5.3.3 Application of Food Safety Regulations at the Bakery

On asking about the food safety regulations, she was unaware of certifications and set of the specified rules to be followed while preparing the consumable food products. As according to her, she was neither asked about it by any government officials or non-government agencies nor was instructed to follow certain rules in her seven years of being in food business.

Besides, she follows the basic food safety rules, like she cleans her shop, working surfaces, and all the utensils (such as; cooking pots, chopping board, knives, and blender jugs), thoroughly in the morning before starting her working day and in the evening after closing her shop. She washes all the washable ingredients such as fruits before using them in product manufacturing, keeps them covered, and stores at cold temperature. She has separate duster cloth pieces for drying the crockery items and furniture for the customers in the shop. The table sheets and other covering material are washed regularly and dried under sun to kill the heat sensitive microbes. Separate garbage bins are used for wet food wastes and other related wastes with coverings to keep the bins' openings closed. Moreover, the water used in food products such as making juice and coffee is sure to be treated for pathogen eradication. She considers the expiry time of the ingredients and products' shelf life to be of vital importance; therefore, she buys only 1–2 kg of the coffee every month to avoid change in taste and aroma over time. Apart from cleanliness of the material and place, she also takes care of personal hygiene by washing her hands with soap before and after preparing the items, after touching surfaces, using toilet, and wears specific clothes and apron which are washed regularly, and keeps her hair covered while handling food material to avoid any physical contaminant find its way into the food.

Concerning the shelf life and product storage, Ms. Thuenprom does not make products in bulk as she uses estimated amount by her experience of approximate number of customer each day since last seven years. Therefore, all the products get sold out on the same day, but if seldom products are unsold, they are distributed in her neighbors or kids rather than store for the next day.

By following above basic practices, she claims that no mishaps have occurred in her business, and no customer has complained till the date about getting sick after consuming her products.

5.3.4 Technology and Logistics at the Bakery

The coffee shop does not use much-advanced technology. The shop has webpage on social media applications such as Facebook Fanpage (Fig. 5.2), where the daily activities and products are advertised. The page also includes the map to find the Coffee Moona shop, along with mobile number. The fan page and the mobile number can be used to place online orders. Moreover, she uses some basic technological tools such as oven, coffee maker, blender, and grinder, which are normally required in bakery.

Although she takes online orders less frequently, if ordered by close friends and relatives (because she cannot bear the expenses of logistics), Ms. Thuenprom uses her family car to deliver the order. She also uses her car to buy the raw material if

Fig. 5.2 Reference to the webpage of the Coffee Moona in Facebook. *Source* Authors

required in bulk, but if she faces a shortage of one or two raw materials, she travels through public transport such as local buses.

5.3.5 Adopted Ways of Marketing and Promotions

Having lesser profit from the business, not many marketing and promotional strategies are used so far, but having more concern about the environmental sustainability issues, Coffee Moona has provided the offer of 5 ฿ discount on bringing personal containers for the beverages and coffee. Earlier, she was providing the customers with the stamp collection cards (also called as reward card), through which a person would get his card stamped every time a beverage is bought, and after collecting nine stamps, the person would get a drink for free. But now she has stopped providing the stamp cards. Moreover, her daughter who studies in Asian Institute of Technology visits her home on every weekend and advertises her mother's products through her personal Facebook page. If any of her friends orders the products, she delivers the products for free on coming back to institute on weekdays.

Ms. Thuenprom plans to give more promotions in the future depending on the budget and income from the business.

5.3.6 Problems Faced in Business

One of the major issues that Ms. Thuenprom faces is the location of the coffee shop. She has built her shop in her home's vicinity which is a residential area due to which the shop is not on the main road or commercial area; therefore, very less people pass by and know about it to stop by and get a drink. Secondly, the area has many shopping superstores such as Big C, Tops, Seven eleven, and there is another coffee shop just across Coffee Moona. All these increase the competition for the customer availability. She also mentioned that she has to look after her home, children, and her mother who is sick; therefore, she cannot focus on her business fully and cannot afford any employee to give her hand in doing the home chores or the work at shop.

This year, the government has made new rules. Those who have their business in their homes (residence) are also needed to pay tax for the place and business, which also increases her worries for the maintenance of the business and expenses. Moreover, if any machine stops working, she has to wait for a long time to get it corrected from the repairing shop as she cannot afford new machines to replace the older impaired ones.

5.3.7 *Planning for the Future*

Keeping in account her schedule, she is thinking about increasing online order and delivery business as more deliveries at the same time will reduce the expenses of logistics. Her most common business targets will be the government organizations and educational institutes where various events are conducted, such as seminars, and workshops. She can deliver full snack boxes containing bakery items and beverages.

Moreover, she is planning to register in various online food delivery applications such as GrabFood and Foodpanda, so that thousands would see her products of people which will increase her chances of getting more online orders from the customers.

Looking at all the problems she is facing currently, we suggested her a few possible solutions such as;

The new tax-paying rule for those working in their own residential areas can increase financial burdens; therefore, they can make a petition requesting the government to withdraw this new rule and get signed by as many as possible business people who work in their vicinities.

Secondly, customers check the ingredients and the product's safety value. The better proof would be the certification from any international or national food safety regulatory authority whose logo on the package acts as evidence for the consumer and environmentally friendly product.

To let people know about her business, she can collaborate with local YouTubers to advertise her shop and products in their videos, highlighting the food safety rules that are followed and the certificate acquired for the practices. Furthermore, as per today's demand, people now believe in what they see for themselves; therefore, selecting random customers to visit the shop's kitchen regarding cleanliness and product quality will help gain customer's trust. Free tasters for the old and new products will allow the customers to find the right taste and by this strategy customers will feel valued and decide to buy any of the products. Similarly, promotions based on various occasions and special days throughout the year will also increase customer's interest in products.

5.4 Sustainability in Food Supply Chain

According to estimations, the world's population will be 34% higher (9.1 billions) than the present population by 2050. With this increase, increase in food production will be needed by 70% of the present day food production. It is also thought that the urbanization will also increase which will reduce the land for the food production. Additionally, water and energy being in limited quantity will be in scarce. Therefore, a sustainable food production system is of vital importance.

Sustainable food production can be defined as the method of production with processing and systems which do not cause any pollution or reduce the amount of pollutants released in the environment, conserve the nonrenewable energy and natural resources as much as possible, having economic efficiency, safer for the workers, consumers, and communities, and do not consume the resources of which the future generations are the rightful owners (users).

In other words, sustainability is the use of resources in amounts that do not surpass earth's capacity of replacing them. In food, food supply security, health, safety, cost-effectiveness, quality, better job and growth opportunities, and simultaneously, environmental sustainability in subject to climate change, biodiversity, soil, and water quality.

The food industry has a greater impact on the environment, as it contributes to climate change, scarcity of water (70% freshwater resources are used by crop and livestock systems with 60% of land), degradation of soil, and biodiversity loss. Therefore, world food production methods should be changed to reduce harsh impacts on the environment and the sufficient production of food to meet the increased population's demands in the future. It is said that crops and animal production and forestry release about 25% of total greenhouse gases emitted globally directly or indirectly [10].

According to European Commission, in this century, fossil fuel use has increased by 12 factors which require extracting 34 times more from natural resources. Among food materials, each person consumes 16 tons every year, out of which 6 tons are wasted and half of this wasted food goes to landfills, due to which sources of minerals, metals, energy, fish, timber, water, fertile soils, clean air, biomass, and biodiversity are under stress. By 2050, demands for food will increase by 70%, 60% of world's ecosystem would already be degraded. If the same way of usage of resources was carried on, we will require resources equal to more than two planets to sustain the living [20].

5.4.1 Highlighted Areas

Sustainable food system can be made possible by modern technologies and knowledge. The actions in this regard include;

- Reduced use of fossil fuels and regulated use of water in production systems.
- Adjusted use of land and avoiding conversion of more land into agriculture.
- Water and energy-efficient food production sites should be designed.
- Monitored use of pesticides and fertilizers to reduce soil and water pollution.
- Reverse the soil erosion and restoration of organic material in the soil.
- Reduction in greenhouse gases emissions.
- Increased biodiversity by adopting farming practices suitable for protection of land, water, and energy sources along with the forests and wetland.
- In case of marine resources, implementation of sustainable fishing practices to restore fish quantity.
- Reduction of pollution due to fertilizers and garbage at coastal areas [1].

5.4.2 *Food Waste*

Food waste has become a serious sustainability global issue. Food waste is defined as the waste or food losses that occur during the food supply chain which include main step as agricultural production, postharvest handling and storage, processing and packaging, distribution, consumption. In annually, food produced around 1.3 billion tons is lost or wasted, one-third of total food produced for human consumption. Food waste can occur at every step of the food supply chain. The waste food of all commodity groups (e.g., fruits and vegetables, fish and seafood, cereal products, meat and dairy, oilseeds and pulses) along the food supply chain is approximately 20–50% [9].

Therefore, people should be concerned about some causes of food waste and know how to solve sustainability issues.

5.5 Solutions

This section presents a few solutions cum strategy for attaining sustainability in food supply chains.

The prevention of food waste is mentioned at all international, national, regional, and local levels. In the international level, the United Nations frameworks a specific aim within the Sustainable Development Goals to reduce food waste volumes reach one-half by 2030. While regional and local levels, the prevention of food waste is gaining interest in food systems sustainability [8]. Most solutions to food waste are important for sustainability in the food supply chain. However, the government and people should collaborate to prevent or reduce food waste in the food supply chain.

5.5.1 *Vertical Farming Toward Managing Energy Waste*

Vertical farming also linked to "indoor farming, urban agriculture, and controlled-environment agriculture" but remains unique. This type of farming can be beneficial for the fresh fruits and vegetables' supply chain, as the production of the commodities can be accomplished in a small area, making it less costly and safer and contributing to the sustainability of the environment. In this type of farming, plants are grown indoors in vertically stacked layers with an artificial environment where temperature, lighting, nutrients, irrigation, and air circulation are controlled and constantly adjusted. The most common type is hydroponics, where the plants are grown in tubes connected to liquid supplies of required nutrients and hormones in fixed proportions mixed with water. Vertical farming is used as a term for a solution to feed the world and repair damaged ecosystems. There are no seasons indoors, making it stress-free due to changing weather conditions and other natural elements that can negatively affect the crops. Depending on the plant species, the efficiency of each floor of plants can be equal to 10–20 traditional soil-based areas. This technology also provides various environmental benefits such as eliminating the need to use fossil fuels as fertilizers and various pesticides, herbicides, weedicides for plowing, seeding, weeding, and harvesting. Urban areas can also be used to produce the crops by taking advantage of vertical farming without needing large lands in traditional farming.

Hence, this technology contributes to sustainable crop production by reducing water and land use, decreased waste, and carbon dioxide emission. The controlled nature reduces agricultural runoff which in traditional farming happens by water carrying byproducts and contaminants to fresh water or oceans. Vertical farming in urban areas can reduce energy usage and fuel emissions due to the transportation of crops from rural areas to urban areas and the reduction of food spoilage. Moreover, in urban areas, the abandoned or unused properties can be used rather than overworked land in rural areas. As by 2030, estimated 60% people will be living in cities where urban agriculture systems can be attractive and beneficial way of feeding growing population [4].

5.5.2 *Sprinkle Spray Irrigation Toward Energy Waste Management*

Spray irrigations are used in large farms for the production of raw material for food supply chains. It has long fixed tubes at one end with water source, from which water flows and is shot out by the spray guns system. Here, the technology also plays its role; for example, Verizon is doing tests where the sprinkler systems are installed in vineyards and the telemetry in the ground to measure moisture content which then turns on the sprinkler system only if there is a need for water. Telemetry is a great example of the internet of things to save wastage of large quantities of water.

Apart from these, soil erosion is one of the problems: the loss of top soil and soil nutrients. This can be due to natural effects of wind and water activity but also human activities such as deforestation and poor land management. The possible solution can be avoiding monoculture practices, i.e., planting different types of crops in the same field, increasing the nutrients in soil, and causing the reduction of pest and crop diseases [4].

5.5.3 *Marine Sources of Food for Energy Waste Management*

Seafood consumption has increased overtime which accompanies the overfishing of specific species of fish and other marine life. Due to the seafood's numerous advantages and nutrients, 29% of marine fish stocks are overfished and 61% are fished at full capacity. Due to overfishing, many fish species are endangered. Therefore, a solution can be fish farming which is also called aquaculture. Farmed fish proves to be a great solution for the seafood supply chains, both fresh seafood and its products as farmed fish is a safe, sustainable, and nutritious source of seafood. It can reduce the burden on overfished marine sources, and it also meets the changing demands of consumers concerning animal welfare and safety of the products. Farmed fish is fresher at the time of purchase as it can be cultivated at any place globally, even in the countries that do not have coastal area and sea as a source of seafood. Secondly, nutrient composition of fish is effected by various factors such as season, diet, location, age, and species. Therefore, nutrient levels differ in marine fish while the feed can influence that of farmed fish. The fish feed can be enhanced by adding microalgae and plant-based nutrients, which results in increased fats in the farmed fish than the wild fish. The fish fats have greater advantages for health and can be used to make various fish oil products. Moreover, human activities and industrialization have caused ocean bodies to be polluted, influencing wild fish and shellfish. The level of contamination depends on the diet of the animal; for example, carnivores such as salmon and tuna might have higher level of contamination. While the diet of seafood animals cannot be controlled which also effects the degree of contamination, the farm fish's diet can be closely monitored and controlled with additional safety provided by the vaccination programs to reduce disease outbreak [13].

5.5.4 Livestock as a Solution Toward Energy Waste Management

Animal products are major source of proteins which contains all essential amino acids but it could be partly replaced by more sustainable sources such as soy and legumes which are protein-rich sources from crops. One such example is the meat analogue produced by vegetarian butchers, in which plant-based meat is prepared with almost similar taste and texture. And such meat replacers are getting consumer's attention gradually.

Moreover, substitutes for animal proteins can also include algae, edible insects, and duckweed. Insects tend to be consumed as a rich source of protein with lesser greenhouse gas emissions. Many insect species are consumed around the globe, with lesser evidence of hazardous effects on health. The level of contamination depends on the methods of production, harvesting, and processing techniques. Insect meals can prove even more environment-friendly by feeding insects on food waste.

Algae (microalgae, macroalgae) have a higher content of proteins, fast-growing, and highly productive than conventional crops. Seaweeds (marine algae) are staple food in Korea and Japan. Furthermore, duckweeds are small marine plants used as feed for domestic animals and mixed in soups salads, etc., in some parts of Asia, a high-quality protein source.

Finally with advanced technology, the meat is produced at the laboratory by culturing stem cells from the animals, which does not require growing a whole animal with expenses of feed for the animals, space, and compromised environmental sustainability. This meat requires lesser space and time, with controlled and known amount of each constituent in the meat. But this technology is still used for research and on laboratory scale, and further advancements and agreement of the consumers are required for it to grow to an industrial scale.

5.5.5 Improvements in Storage Warehouses

5.5.5.1 LED Lighting

An enormous amount of cool temperature and light is required for the storage of products and raw material. LED lighting can be used in cold storage, which is energy efficient and generates less heat.

The Yusen logistics in New Jersey has shifted to LED, which has reduced their electricity use and expenses by 60%, accounting for 1 million kilowatt-hours (kWh) per year. Another company, Atlanta-based Americold logistics, which is temperature-controlled warehousing and Transportation Company for perishable goods, has installed motion-sensing LEDs with fast-closing doors. The LED lighting system activates as needed by this system and switches off automatically when the personnel leaves. That saves 14 million kWh per year [1].

5.5.5.2 Solar Power and Turbine Generators

The Americold Company uses solar power generation at Massachusetts facility which generates 500,000 kWh of clean energy and turbine generators at Salina, California, that generates up to 600,000 kWh of energy per year. And according to the company, they have saved 320 million kWh since 2010, which means 264 million pounds of CO_2 was not released into the atmosphere [13].

5.6 Conclusion

The supply chain connects the members working together and bringing a product to the customer. The food supply chain is the network for a food product from its producer to supplier and distributor till it reaches its consumer. The food industry has a major influence on most of the Southeast Asian countries' economies. Being ranked as one of the top exporters of one or more food products, the ASEAN is trying to improve their food supply chains at each step by using better technology, food safety systems, packaging, etc. Although some of the member countries still have traditional supply chain systems, the Association of Southeast Asian Nations' newer rules and regulations are helping these countries adopt modern and better supply chain practices. Moreover, as the resources are limited for the growing population, adaptation to more sustainable supply chain systems is important. And sustainability in the food supply chain is the requirement of the Southeast Asian food industry and every country around the world. The major issues include energy waste, food waste, and pollution of the environment and its components. In this scenario, wiser ways of resource utilization are required. In this context, many innovative solutions are brought in and some need testing for their feasibility. Newer rules and regulations are set, but for implementing the solutions and laws, each country's government and the public need to collaborate and work together.

References

1. Beddington JR (2011) The future of food and farming: challenges and choices for global sustainability [final project report of the UK government foresight global food and farming futures]. Government Office for Science
2. Bukeviciute L, Dierx A, Ilzkovitz F (2009) The functioning of the food supply chain and its effect on food prices in the European Union. European Commission Directorate-General for Economic and Financial Affairs Publications, Brussels, pp 4–5
3. Buurma J, Saranark J (2006) Supply-chain development for fresh fruits and vegetables in Thailand. In: Agro-food chains and networks for development. Kluwer/Springer, Dordrecht, The Netherlands, pp 119–127
4. Chabada L, Dreyer HC, Romsdal A, Powell DJ (2013) Sustainable food supply chains: towards a framework for waste identification. In: IFIP advances in information and communication technology. http://doi.org/10.1007/978-3-642-40352-1_27
5. Chakraborty A, Mandal P (2014) Understanding challenges of supply chain sustainability in Asia. Int J Process Manag Benchmarking 4(1):51–68
6. Chonhenchob V, Singh SP (2005) Packaging performance comparison for distribution and export of papaya fruit. Packag Technol Sci Int J 18(3):125–131
7. Dani S (2015) Food supply chain management and logistics: from farm to fork. Kogan Page Limited, UK, pp 1–15
8. Diaz-Ruiz R, Costa-Font M, López-i-Gelats F, Gil JM (2019) Food waste prevention along the food supply chain: a multi-actor approach to identify effective solutions. Resour Conserv Recycl 149:249–260
9. Gustavsson J, Cederberg C, Sonesson U, Emanuelsson A (2013) The methodology of the FAO study: "global food losses and food waste—extent, causes and prevention". In: FAO, 2011. SIK—The Swedish Institute for Food and Biotechnology, Lund, pp 8–13
10. Hochfelder B (2017) Sustainability in the food supply chain. Food Logistics
11. Hu FB (2002) Dietary pattern analysis: a new direction in nutritional epidemiology. Curr Opin Lipidol 13(1):3–9
12. Hugos M (2011) Essentials of supply chain management, 3rd edn. Wiley, Hoboken, pp 1–38
13. Li D, Wang X, Manzini R (2014) Sustainable food supply chain management. Int J Prod Econ 152:1–6
14. Lu D (2011) Fundamentals of food supply chain management. Dawei Lu & Ventus Publishing ApS, pp 8–27
15. Noomhorm A, Ahmad I (2008) Food supply chain management and food safety: South & East-Asia scenario. Agric Inf Res 17(4):131–136
16. Othman NM (2007) Food safety in Southeast Asia: challenges facing the region. Asian J Agric Dev 4(1362-2016-107682):83–92

17. Otto B, Sentana M (2013) Indonesian capital's infrastructure strained by flooding. Wall Street J. Jan 20 [Online]. http://online.wsj.com/article/SB10001424127887324624404578253380604853110.html
18. Pingali P (2007) Westernization of Asian diets and the transformation of food systems: implications for research and policy. Food Policy 32(3):281–298
19. Rao P, Holt D (2005) Do green supply chains lead to competitiveness and economic performance? Int J Oper Prod Manag 25(9):898–916
20. Roadmap to a Resource Efficient Europe (2011) Communication from the Commission to the European Parliament, the Council, the European Economic and Social Committee, and the Committee of the Regions. Roadmap to a Resource Efficient Europe, Brussels, p 20
21. Rolle RS (2011) Packaging in fresh produce supply chains in Southeast Asia. FAO Regional Office for Asia and the Pacific, Bangkok. ISBN 978-92-5-106998-1
22. Scott W, Oldfield C (2004) The nine basic rules of a successful supply chain. Waer Systems, Ltd., UK
23. Seo K, Lee C (2007, May) Development and prospecting of agricultural traceability system in Korea. In: International symposium on traceability for food safety. RDA, Korea, p 154
24. Teng P, Escaler M (2010) The case for urban food security: a Singapore perspective. Centre for Non-Traditional Security Studies, S. Rajaratnam School of International Studies
25. Wong LC (2007) The role of supply chains and international networks in enhancing the agricultural sector's competitiveness in Cambodia, Laos, Myanmar, and Vietnam (CLMV). Asian J Agric Dev 4(1362-2016-107657):1–10

Chapter 6
Traceability and Transportation Issues in the Food Supply Chain

Akshay Patidar, Monica Sharma, Rajeev Agrawal, and Kuldip Singh Sangwan

Abstract The food supply chain (FSC) is complex, immense and crucial to human beings. Disruptions, fragmentation, poor product traceability, improper product flows, food contamination, food recall, etc., are some of the prevailing issues in the food supply chain. Many cases of *E. Coli*, hog contamination and halal meat contamination are also evident. This creates a strong need to identify, understand and resolve the issues, leading to an efficient and responsive food supply chain. This study identifies the various issues in the food supply chain and thereafter focuses on traceability and transportation-related issues. The authors have tried to conduct a review of the issues in the food supply chain, especially traceability, transportation and solutions under the purview of supply chain drivers. The study can help future researchers identify and understand an issue and develop a feasible solution for it.

Keywords Traceability · Transportation · Food supply chain

6.1 Introduction

The prime objective of a business is to fulfil a customer's need. To do so, an organisation has to procure raw materials, process them, produce products and ultimately, market the product, ensuring its availability. Here exists the concept of supply chain:

A. Patidar (✉) · M. Sharma
Department of Management Studies, Malaviya National Institute of Technology, Jaipur, India

M. Sharma
e-mail: msharma.dms@mnit.ac.in

M. Sharma · R. Agrawal
Department of Mechanical Engineering, Malaviya National Institute of Technology, Jaipur, India
e-mail: ragrawal.mech@mnit.ac.in

K. S. Sangwan
Department of Mechanical Engineering, Birla Institute of Technology and Science, Pilani, India
e-mail: kss@pilani.bits-pilani.ac.in

R. S Mor et al. (eds.), *Operations and Supply Chain Management in the Food Industry*, Lecture Notes in Management and Industrial Engineering,
https://doi.org/10.1007/978-981-16-5555-5_6

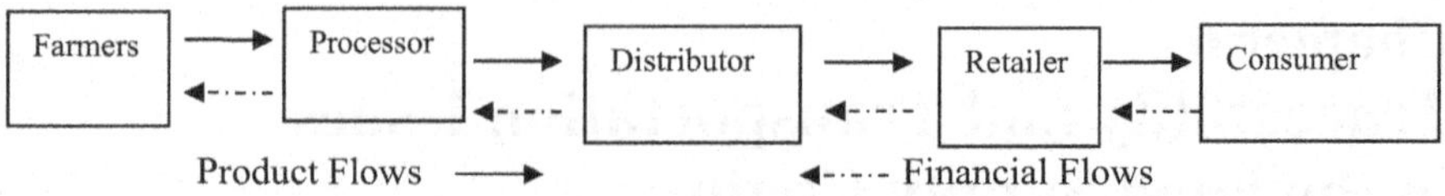

Fig. 6.1 Typical food supply chain. *Source* Author

a customer demands something; on his demand, a manufacturer procures raw material and manufactures the product as per the customer's specifications and sells it to them. A network of all these integrated operations is called a supply chain. Typically, when all the stakeholders come together with a shared objective or goal, it becomes a supply chain. Keith Oliver in 1982 coined the term "Supply Chain Management", where there is a strong requirement for managing and ensuring the smooth flow of goods, information and finance [47]. Based on the nature of the product, the existence of the supply chain is defined. A typical food supply chain is shown in Fig. 6.1, where the flow of products and money can be seen along with the stakeholders of the supply chain. Information flow is not shown in the figure and is bidirectional.

These flows of a supply chain are facilitated by several pulls and pushes. In a food supply chain, farmers or processors, after processing food, push the food to consumers, who, using their demand, pull the product from the processor or the producer. This push and pull facilitates the supply chain, but when managed improperly creates some issues like the bullwhip effect. To overcome this problem, one has to manage the information flow effectively. It is important because it creates efficiency throughout the supply chain and should be bidirectional in the case of a food supply chain. The information flow facilitates decision-making by sharing information with all the stakeholders across the supply chain.

Supply chain in advanced or developed countries is categorised into various sectors such as agriculture, horticulture, fisheries and aquaculture. These fall in the category of primary producers, whereas manufacturers who process the food (and make it ready to cook or eat) and the various packaging organisations are intermediaries. Wholesalers, retailers and caterers are the final part of the food supply chain [61].

Across the entire supply chain, value is added with the change in ownership. Food items can reach consumers from any of these stages, i.e. fresh produce from farmers, caterers and manufacturers. This movement is supported by the logistics service provider, who can be in-house or a 3PL service provider. Much fragmentation or primitiveness can be seen in developing countries like India.

Professor N. Viswanadham, in his special feature, discussed Indian leadership in the food supply chain. He said that to become a leader in food supply chain, India needs to focus on introducing supply chain expertise by embracing efficient customer response (ECR), which was introduced by the USA in the early nineties and is now globally accepted in grocery supply chain [67]. This is a strategy that leads all companies to work together to serve the customer better and at a lower cost. Also, it emphasises collaborative planning, anticipation and replenishment. The customer gets availability and choice as benefits from the distributors and retailers.

Professor Viswanadham discussed the need for standards in food supply chain: like food standards that are related to content and processing, or related to logistics and IT systems that will enhance the compatibility, interoperability and ultimately, productivity. Further, he discussed multiple business models, where a customer is served in many online and offline ways by the food processing industry. International customers must be served through food malls, and for effective decision-making, their lead time, service time and demand forecasting need to be focused on. He further discussed streamlining processes and several prevailing regulatory measures of the state and central governments. Streamlining these would be highly productive. Further, he proposed that research needs to be initiated for developing indigenous packaging, machining and laboratories for food processing and storing. Training is an important part and needs to be focused on to make India a leader in the food supply chain. For this, all the stakeholders of the food supply chain have to be coached, counselled and mentored. Apart from these generalised issues and their remedies, several other issues need to be focused on.

This study is an effort to review the prevailing issues in the food supply chain, especially traceability and transportation issues. Section 2 discusses the issues. Section 3 focuses on the traceability issues, while Sect. 4 discusses transportation issues. A few solutions are discussed in Sect. 5, and conclusions are listed in Sect. 6.

6.2 Issues in the Food Supply Chain

The food supply chain is facing many roadblocks. It is important to make it resilient to these roadblocks. Whether climate change happens, or a disaster or epidemic occurs, it leads to disruption and directly affects the food supply chain. Apart from these, many other issues like creating provenances (which is seeing high demand from consumers these days), traceability and transportation issues are also seen in the food supply chain. These need to be addressed. For this, *understanding*, the issues is of prime importance.

6.2.1 Climate Change

Author Steve Banker, in his articles published in *Forbes*, identified the complexity of the food supply chain and addressed the impact of climate change on it [60]. Further, he discussed the report by the Intergovernmental Panel on Climate Change (IPCC) and found that major climate change affects food supply. Researchers estimated that, for every 1 degree increase of temperature on the planet, crop yield decreases by 10%. The latest estimates say that reliance on fossil fuels needs to be reduced, thereby reducing emissions. This will halt or reduce the increase in temperature. Hence, the author suggested that emissions be curbed, thereby reducing the climate change effect so that the food supply chain faces fewer disruptions and can provide an uninterrupted supply.

6.2.2 Natural Disasters

Extreme weather events are common these days. During the aftermath of a disaster, or the response or recovery phase of disaster management, we see the importance of providing aid to the affected area. Many of us have seen people trying to stock up supplies before a storm hits them. To maintain these supplies without interruption, farmers need to conduct proper evacuation of their animals. In North Carolina, hog farming is a major industry worth $2 billion. Hurricane Florence hit this region and directly affected hog farming at Sampson and Duplin. Overall, more than 6000 hogs perished in the storm, leading to a higher price of pork due to limited supply [33]. To deal with such risks, control towers are being developed, these are commonly known as holistic control towers.

6.2.3 Epidemics

Pathogens, especially airborne pathogens, are a common threat to the food supply chain. Back in 2017, a chicken farm associated with Tyson Foods found that a flock of chicken having 73,500 birds had been contaminated. The flock was being used for breeding purposes. As a result, the entire flock had to be destroyed and farms within six miles of the contamination were quarantined [65]. The implications of the recent COVID-19 pandemic are widely known. The effect of the pandemic on the food supply chain and the related disruptions are also well known. Social distancing norms and lockdowns interrupted the food supply chain, leading to wastage of food supply. Increased demand from consumers, along with increased prices, shocked the entire chain. For food-related organisations, this resulted in an increased struggle to meet consumer demand, which led to increased allocation. But, despite such allocation, supply chain disruption leads to a risk of losing market share.

6.2.4 Creating Provenance

Increased awareness among consumers led food companies to create provenance of their supply of food ingredients. This made large organisations like Unilever, Nestle, etc., modify their procurement practices. The consumer wants to know not only the ingredients but also the impact of the sourcing decision on the environment. For example, with the increased usage of palm oil, sourcing of palm oil is a major concern due to the deforestation of tropical trees. Many large organisations like Mondelez publish online data that consumers can use to get details of the farm or plantation. Consumers can go online and view details regarding where cocoa is procured from and grown. Therefore, stakeholders in the food supply chain are focusing on creating provenance and enhancing supply chain visibility.

6.2.5 Traceability Issues

Tracking and tracing are major concerns of food supply chain managers these days. Because of the increased demand for creating provenance across the supply chain, stakeholders are also demanding that track-and-trace issues be addressed. Creating a sound traceability system is the need of the hour. Food companies are focusing on developing a tracking and tracing system to provide insight to consumers regarding the product's movement through the supply chain. Major traceability issues include food crisis management like food recalls, food quality and food safety, food fraud prevention and most importantly creating traceability of bulk products like grains, cereals, etc. [3]. Food identity preservation and counterfeit products are other issues related to traceability, where the major focus is on food preservation. In the case of identity preservation, a commodity must not lose its identity. Also, counterfeit products should be identified while tracing the product during movement.

6.2.6 Transportation Issues

Transportation is an integral part of any supply chain. It smoothens the most important flow, i.e. the flow of goods. Maintaining these flows caters to many related issues, such as lesser than truckload (LTL) shipping, refrigeration or cold chain facilities for perishable products, on-time deliveries to reduce the chargebacks in case of promised delivery, and handling issues that can lead to contamination of the food products [34]. Refrigerated chain or cold chain is one of the important issues that not only affects the food supply chain but also plays a major role in the healthcare supply chain. Increased cost of equipment and less developed infrastructure are the prime factors because of which the cold chain remains less developed. But once it is developed, it will lead to the resolution of major issues in the food supply chain and healthcare supply chain [45]. These issues need to be addressed to create an efficient and responsive FSC that can contribute to increased customer satisfaction and a less fragmented food supply chain.

6.3 Traceability Issues in FSC

The definition of traceability and traceability system found in the literature seems to be very broad [48], but the common thing in it is oriented towards the capability to guarantee the tracking and tracing ability of a food supply chain (FSC), in relation to its products. As per European Commission (EC) Regulation No. 178/2002, traceability implies "the ability to trace and follow a food, feed, food-producing animal or substance intended to be, or expected to be, incorporated into a food or feed, through all stages of production, processing, and distribution" [30]. The Codex Alimentarius Commission (2206) describes traceability as "the ability to follow the movement of a food through specified stages of production, processing, and distribution" [11].

According to the author [18], tracking can be defined as the ability to follow the path of a product along the supply chain, whereas tracing refers to identifying the product's origin and its characteristics by referring to records in the supply chain. Due to reasons like mandatory regulations, international standards and certification requirements, development of marketing strategies, creating provenance for product origin, maintaining quality standards and safeguarding from sanitary outbreaks, tracing the history of a food product related to its displacement is essential for modern companies.

For an individual firm, the traceability value is related to its differentiating ability and selling the product when a competitor has provided an unsafe product to the market [55]. For an industry, it is important to safeguard its reputation and market share by differentiating the product from the problem. And for a consumer, it is about removing contamination before fulfilling family needs. On a major scale, consumer interest is related to human health.

As per ISO 2005 and 2007, traceability system (TS) is the integration of data and operations, maintaining the desired information pertaining to a product and its components through the entire value chain. TS maintains records and follows the trail of a product or material: it enters the supply chain through suppliers, is processed and is distributed as an end-product.

A basic system has these characteristics:

- Batch or unit identification.
- Information storing with the movement of batches/units.
- A system linking the identification and product movement with relevant traceability information as the product moves to the next stage.

Among the different kinds of food traceability based on how traceability is being created and what is it all about, the following are mainly discussed [36]:

- Conventional traceability: the traditional way of creating traceability using a labelling system, and its management for processed foods, but in batches.
- Genetic traceability: creating traceability where a product's genetic composition is determined to detect fraud and valorise production.
- Geographical traceability: tracing the point of origin and creating provenance in the system.

Creating a sound traceability system is becoming the need of the hour. But it has several problems. Disease outbreaks, poor supply chain visibility, complex global supply chains, etc., are factors that affect the creation of a traceability system in the food supply chain. Further, this section focuses on the various issues with respect to traceability systems that can have a significant impact on the food supply chain and its performance.

6.3.1 Food Crisis Management and Food Recalls

The mechanism of food recalls and crisis management is a significantly underdeveloped field [12]. Recall may be defined as an action to call back such food products from the food chain as may pose a threat to public health or violate the country's law, including a product that is in the customer's possession. For an efficient recall, the architecture of food safety systems needs to be developed, as it affects the organisation's ability to recall. Recall is done in public interest [10]. Besides product recall, product withdrawal is also done in some cases where there may be public health and safety issues, or product issues. It can be done in two ways:

- When the product has quality defects or labelling irregularities.
- As a precautionary measure, pending further investigation, and where a threat to public health and safety is established.

Food recall can be done due to many reasons [1]. One such reason is incorrect food packaging. This may include incorrect packaging, undisclosed allergens, misleading claims and incorrect ingredient labels. Another reason for recall is that the food item contains biological, physical or chemical hazards. For example, a high level of *clostridium botulinum* bacteria, which causes botulism, can lead to the recall of canned goods. Similarly, metal shavings from machines getting into a product can lead to a physical hazard. And so, the food is discarded.

Another example of food recall is farm produce that has been cross-contaminated from animal manure and runoff. This can lead to *E. Coli* spreading to fruits and vegetables, as happened in the romaine lettuce outbreak in Arizona. Pathogen-contaminated products are highly damaging because they affect all consumers. According to USDA's report, Listeria, *E. Coli* and Salmonella led to 17 food recalls in 2015.

6.3.2 Traceability of Bulk Products

Many industries use bulk ingredients in the form of liquid, powders, crystals or grains stored in huge silos that rarely get emptied. Hence, several lots are kept in the same silos, and continuous consumption takes place. Traceability becomes very difficult in the common case where bulk products are used because of a lack of knowledge about the lot's composition [14]. Similarly, fruits, vegetables or grains in the food supply chain are some of the bulk commodities, and tracing them is a big challenge if they get piled from different sources of different kinds and at different periods. Many authors have developed models to create a traceability system for such problems. The author [59] discussed operational techniques for implementing traceability in bulk product supply chains, and a system approach was used to develop a framework where a relational database model was developed to facilitate internal traceability. It was suggested that this technique could mitigate bulk grain handling. Similarly, authors Li [39], Liang et al. [40, 41] proposed a set of solutions, like inserting pill-sized food-grade tracer particles during harvest. These tracers can provide information related to the point of origin of the product and are edible, such as sugar and cellulose. Specialised food-grade ink is used in printers to print barcodes. The authors used dynamic simulation to model the changeovers of liquid in a pipe for bulk liquid products [54]. The author proposed a traceability model for the raw Brazilian sugar supply chain using RFID and WSN [29]. Apart from this, a lot of research is being done to develop a traceability system for products like soybean, corn, etc. A study was conducted and a traceability system developed using item-level radio frequency identification. This was applied in the dairy industry to trace milk as a product [6]. Another interesting study discussed the breadth, depth and precision of a traceability system [32]. The author also discussed the benefit of implementing a traceability system in private organisations. Further, they stressed upon UPC that most of the products have a barcode that is used to determine consumption pattern and sales pattern, and now, these barcodes can be used to track numerous product attributes. Another study discussed employing various chromatographic techniques such as gas–liquid chromatography, high-performance liquid chromatography and electrophoretic methods for authenticity tests and traceability of vegetable oils and dairy products [16].

Similarly, another study focused on the labelling and traceability of genetically modified organisms and discussed DNA-based detection methods and protein-based methods. Finally, the author discussed that the feasibility of traceability systems depends on several factors, including identifiers for GM products, detection methods, level of contamination and financial costs [46]. Another Indian study discusses the Indian perspective on food traceability and the initiatives taken by APEDA and its association with GS1 India in the form of Anarnet, Peanut.net, Meat.net and Grapenet and several ICT initiatives working in many states of India [19]. Also, the author discussed several factors affecting the traceability system, such as restrictive government, marketing standardisation, insecure policies and unstable actions for food safety, unorganised infrastructure, the absence of cold chain facilities and small local stores and inadequate agricultural practices, with a large number of small and medium industries and farmers.

6.3.3 Quality and Identity Preservation Concerns

The development of technologies like RFID tags has provided new and interesting opportunities to FSC managers. This technology has inbuilt sensors for temperature, humidity, etc., and can transmit the collected data along with the item's identity. Using this, a traceability system develops the ability to capture information about the item's identity and its attributes or properties like humidity or temperature history, thus providing a complete description and status of the FSC to the manager. This widens the scope of dynamic optimal planning, like route planning for perishable goods according to their state of perishability. Some authors have proposed a method to improve the ability to define an optimal target temperature for multi-commodity refrigerated storage [4]. Further, the author conducted a simulation-based experiment with a wireless sensor network (WSN) and found that sensor-based methods for real-time quality monitoring and assessment are better than traditional visual assessment methods. Research in 2014 developed a model to analyse quality control in food supply chain management, performed case studies and found that the tier in a food supply chain is directly proportional to distortion in the supply chain [13]. Food quality distortion can be removed through effective and efficient pricing and regulatory approaches. The authors in 2014 conducted a case study in the red wine industry about logistics data to assure the quality in a sustainable food supply chain [62]. Another author developed an improved traceability system model for food quality assurance and evaluation based on fuzzy classification and neural networks that not only provides efficient traceability but also helps in the evaluation of traceability system as a basis for visualising the quality management process [69].

Adding to it, due to the increased need for product differentiation, the need for identity preservation arises significantly because it adds economic value to the product. Bennet, in 2009, defined identity preservation as the ability to maintain particular traits or attributes [7]. Among these, country of origin, safety-related traits, GMO, halal, allergen contamination, animal welfare, low carbon footprints, etc., can be listed. These attributes are not directly related to quality characteristics, but affect the product value-addition recognised by consumers. However, consumers cannot identify these directly, and so, they need a guarantee through certifications. A well-known study in 2016 identified identity preservation in international feed supply chains and focused on non-genetically modified crops [66]. The author identified three steps, i.e. product management at the origin, product transportation and product processing, and suggested that agricultural products should maintain their identity at all these three levels of an FSC.

A recent study discussed food traceability from a consumer's perspective and answered a research question as to whether the variables characterising Poland's food product consumers—like gender, age, education and financial status—affect food traceability aspects such as identification of the producer, food features' importance, the importance of the information given on food product packaging, and the influence of the shopping place and frequency on tracing the food origin. By conducting empirical research on 500 respondents of Poland, the author found that creating provenance (i.e. point of origin) mattered to consumers, irrespective of the conditions that were analysed [68]. Food origin gained the maximum attention of consumers, and the availability of information about producers was considered necessary.

Identity protection is also used to separate lots based on their specificity of traits and to understand if a particular known composition is being used to enhance the resulting composition blend [42], for example availability of protein in flour and of ethanol in wine. In cases like fortified food, diet food, baby food and industrial products, some components have to be present more than in traditional products.

6.3.4 Fraud Prevention and Anti-counterfeit Concerns

Frauds and fakes are increasing with the increased complication of the FSC and need to be addressed. Mainly, these instances are evident in high-end products like wine, cheese and caviar and result in economic and reputation losses. Implementing traceability will be useful for developing an effective tracking- and tracing-based system that can store the history of product movement along with the point of origin and processing history [50]. With the help of overt and covert technologies, this problem can be resolved. The most promising technology for this kind of system is RFID because of its wide usage [17].

But the problem lies in the management of this technology. The data stored in the central database and the cost of managing that database are comparatively high. Besides this, other technologies like blockchain and cryptography can be useful for the system because of their immutability feature. Data can be stored, but it is very hard to manipulate the information. Also, it is easy to track the information back, as it is shown in the form of transaction history and is decentralised, even though centralisation makes the system more transparent. Many studies have been carried out on creating traceability systems to address the fraud prevention and counterfeit concerns.

The authors in 2015 developed and published a rapid quantitative detection method for onsite food fraud analysis and focused on spectrometry- and spectroscopy-based approaches. They used predictive computing, along with sensor and Internet of Things technology, after the commendable development in computational and information sciences [22]. Some other researchers discussed mitigation measures using traceability systems and data analysis to focus on fraud prevention rather than on fraud detection [44]. The authors in 2014 focused on halal FSC and its integrity, as the halal concept is an object of concern these days [73]. The halal status of a food product needs to be proved to consumers because of their awareness of this concept. Hence, the author focused on halal certifications, halal standards, halal traceability and trust across the FSC members and developed a conceptual framework for investigating halal integrity with the food supply chain. Further, in 2016, a PPP model was stressed upon for food fraud inhibition. Steps necessary to implement country-wide or industry-wide vulnerability assessment were discussed, and a food fraud prevention strategy that is focused on vulnerability assessment and initial screening was proposed [56].

If an active consumer can reach a database using handheld devices like mobile phones, the effectiveness of the traceability system and the trust in the food supply chain will be enhanced. Most of the traceability and anti-counterfeit systems are based on information collected at various levels of the food supply chain. And, as the traceability costs are already included in other purposes of the supply chain, track-and-trace anti-counterfeit systems can reduce the price while protecting food from frauds and fakes, without losing competitiveness [57].

6.4 Transportation Issues in FSC

Transportation plays a vital role in the supply chain. The supply chain focuses on and maintains product flows, information flows and financial flows, and transportation or logistics provides an integral solution for facilitating these flows.

The transportation system contributes to maintaining the flow of goods from farmer to processor and then, to the distributor and retailer in case of agri-food supply chain (AFSC) [63]. Another mode of product flows can be from the farmer to the wholesaler and then to the retailer and consumer in the food grain supply chain (FGSC). In some cases, the cold chain also plays a vital role and needs a temperature-controlled supply chain for perishable products. Refrigerated transportation is needed in the fruits and vegetables supply chain (F&V SC).

Every industry has its own logistics and transportation issues. These can arise from a multitude of sources, such as product type, model, receiver and market conditions. Professionals at Zipline Logistics discuss the unique issues of food and beverage supply chain transportation [24]. These issues need to be managed for effective and efficient food supply chain management.

6.4.1 On-time Delivery (OTD) and Retailer Chargebacks

On-time delivery (OTD) is a challenge in food transportation, as most of the manufacturers or retailers specify a date for delivery—commonly known as "must arrive by dates" (MABD). If these dates are missed by a transport service provider, it leads to a penalty. In an example cited by the professionals of Zipline Logistics [24], Roger charges a flat $500 for late orders and Wal-Mart takes 3% of the purchase price on late cases under its "on-time in full" (OTIF) program. If the manufacturers are unable to meet the timelines as specified by customers, a brand can lose a huge amount of money in a quarter and can even lose a buyer relationship.

To maintain the relationship with customers, many of the retailers are setting strict parameters about their order management. Many retail stores are working on pallet configuration, appointment scheduling and confirmation processes, operating hours and their preferred carriers. Researchers in 2011 conducted a study on cluster-building and logistics network integration of local FSCs to enhance the efficiency of logistics across the potential markets, along with reduced environmental impact. They developed a coordinated distribution system for local food producers and enhanced traceability for consumers [8].

6.4.2 Refrigerated Shipping

Many industries dealing with goods of a perishable nature prefer refrigerated shipping. Refrigerated shipping is nothing but temperature-controlled shipping. It constitutes a major part of the cold chain. Generally, temperature varies from 0 to 4 °C. Initially, the cold chain was used only for vaccines or medicines, but nowadays, it is being used for almost every product that is perishable. Demand for such temperature-controlling equipment rose significantly due to consumers' preference for fresh and less-processed products.

Two reasons the supply chain is seeing little improvement are the higher cost of ownership and a challenging regulatory environment [72]. It directly shows the tighter capacity and increased prices when compared with traditional truckloads. Availability of these refrigerated vehicles is a challenge, also because of the seasonality. They are much-required during prime production season, but are less available on the spot. But during the off-season, the availability of these vehicles is high. This affects not only their capacity utilisation, i.e. operational capability, but also their operating cost i.e. financial capability. Another author in Chap. 16 of *Post-harvest Handling* discussed the fresh produce value chain from field to home, along with refrigeration and other supporting technologies and focussed on increased shelf life and the extension of perishable produce storage [21].

Researchers discussed the possibility of global trading of perishable goods using the creation of environmental control and refrigeration during maritime transportation [23]. Further, the author focused on the mean rate of energy consumption of the systems and discussed greenhouse gases and carbon footprints during the transportation of apples from New Zealand to the UK and their long-term storage. Researchers also compared a refrigerated container versus bulk in the banana cold chain [2]. They conducted a two-fold analysis where, initially, they described the demand and supply of refrigerated products and reefer capacity, respectively, utilising containership fleet and by conventional reefer fleet. Secondly, a trade case of banana was evaluated and a cost analysis carried out for each stakeholder across the banana cold chain. Other authors, in Chap. 18 of *Delivering Performance in Food Supply Chains*, discussed the significance of keeping a controlled temperature during transportation, for the safety of microbes and the improved quality of food [35]. Further, the authors discussed the levels of the cold chain, i.e. chilling, freezing, storage, transportation and the shop-floor level. Energy reduction and environmental issues were also discussed, and finally, some factors that producers should consider while specifying refrigeration systems for the FSC were mentioned.

6.4.3 Handling and Contamination

Handling and contamination are directly connected to food safety and food quality. To maintain safety and quality, several measures have to be considered by the logistics partner. Cleaning the loading space is one of the important dimensions described by the Zipline professionals. Improper cleaning leads to infusion into the food packages that are being shipped, which can contaminate them. Trailers often get rejected for the same reason, and this leads to a delayed timeline, affecting the efficiency and responsiveness of the entire supply chain. Another issue related to food safety and food quality is the over-handling of freight [51]. This problem is evident in snacks-based products and bulk ingredients, where bags and totes are fragile and can get ripped during movement. Chances of breakage are high during loading and unloading.

Damaged packaging or holes in the package leads to a ruined product. This shows a very low product safety. Hence, it was proposed by industry professionals that the modes of transportation be chosen with very few transfer points to maintain product safety during handling. Further, many studies have been conducted on the above-mentioned problem, focusing on the risks of halal food product contamination during transportation [70]. Further, the authors discussed the food hygiene regulation 2009, ISO 9001, Malaysia Standard of MSI500:2009 and MS2400:2010 that cover food production, transportation, warehousing and retailing in Malaysia. Another study characterises risks and controls for food transportation safety, focusing on various types of contamination, like biological and physical, during storing and transporting [52]. Further, the authors elaborated on five food safety hazards that carry the greatest concern, i.e. lack of security, improper handling practices, inadequate loading and unloading practices, contamination across stock and less control of temperature [37]. The analysis also suggests that organisational risk factors and transiting factors explain the relationship among food safety-related hazards. The author's analysis further suggested some measures to mitigate the risk of food damages. These include creating awareness among and training employees, frequent review of records by management and effective communication among stakeholders of the entire food supply chain.

6.4.4 Less Than Truck Load (LTL) Shipping

Less than truckload shipping is beneficial to those who are not able to fill a full truck, since they do not have enough freight. The traditional LTL model requires pricing up of freight and unloading at a central facility and then, reloading for delivery, which leads to distortion due to extra transfer of product. A factor that can resolve the LTL shipping issue but create another issue is the loading of a large mix of products. But then, a threat of contamination is created, thereby making LTL a difficult mode for consumer brands. Many authors, through their research, have discussed this issue, along with load securement techniques and packaging methods as a mitigation tool for reducing the risk of damage and injury during shipment by a truck or container.

Also, they have discussed proper loading methods, along with food pallets and unitisation methods to reduce subsequent injury during loading and overcoming penalties and other related matters. Further, the authors have discussed requirements to meet Federal Motor Carrier Safety Regulations (FMCSR) [53]. Another study, conducted for a scheduling problem for cross-docking with loading and unloading constraints based on an improved particle swarm optimisation algorithm, compared the improved particle swarm optimisation with PSO and GLNPSO and found that improved PSO outperforms all the algorithms, particularly when there are more products, docks and trucks [71]. The author performed statistical tests to find out the statistical significance of the difference in performance. Another author discussed the same scheduling problem in "less than truckload" industries [43].

6.5 Resolving the Issues of the FSC

The FSC is growing rapidly and so are stakeholders' expectations. For instance, consumers want fresh and exotic food on their plates throughout the year. This expectation has led supply chain managers to extend their verticals across geographies and to more stakeholders, leading to a longer and complicated supply chain.

We frequently come across several issues in the food supply chain. Here are some suggestions to resolve them.

6.5.1 Lack of Traceability

Lack of traceability and transparency develops a huge grey area in the supply chain, which can expose the entire supply chain to several kinds of risks. This lack (of tracking and tracing issues) in the FSC arises mainly because of the use of outdated or traditional systems [9]. This introduces errors and delays in information sharing and lowers the supply chain visibility.

A technological solution to these issues can be blockchain technology. Many researchers and authors have described this technology as an important one for creating tracking, tracing and transparency in the FSC [31, 49].

Blockchain technology is a distributed-ledger technology, where data are stored in a distributed manner and the information is shared across the network, leading to better supply chain visibility. As a benefit, users can view the information on a real-time basis without having the right to alter information. Another advantage of this technology is that once the information is circulated through the network, no one can alter the information: hacking, cracking, manipulating and corrupting information are a time-consuming and costly affair. The essence of this technology is creating transparency, traceability and trust using the power of decentralisation and a consensus mechanism that not only empowers the stakeholders but also uses the consensus to store the information. The information is fetched from a decentralised database that stores information after obtaining the consensus of the key persons having the right authority [5].

Many researchers are developing a farm-to-fork model that can trace even a single bit of information across the FSC.

Technology has the power to deliver transparency, traceability and trust that have been the requirements of the food industry for a long time. Because of this unalterable data, a system developed can share information pertaining to the point of origin, ingredients and current state of the product with all the producers, suppliers, distributors, retailers and consumers [58].

6.5.2 Product Safety and Quality

Manufacturers these days face pressure to deliver safe and high-quality products. Commonly identified causes for these issues are storage and warehousing practices, transportation delays, improper weather conditions and industrial sabotage [25]. These safety and quality issues sometimes lead to product recalls, which result in huge costs and can result in damage to the brand equity also.

For manufacturers who want to create product safety and product quality, this entire process starts with the selection of raw materials and the implementation of appropriate production methods as per the standards, along with testing them and creating provenance [27]. Testing entails the selection of an accredited laboratory that practises recent and advanced collection and analysis techniques to ensure accurate results and reliability of quality assurance. Besides this, a trustworthy logistics partner needs to be identified. They must be experienced in handling food products and must have a good track record and reputation.

People from academia and industry are continuously trying to improve industry standards with the help of research and development. Many authors have published their research in reputed journals directly related to food safety and food security. Researchers are also developing models as per international standards to create international acceptance of the food product. This will not only contribute to the improvement of the industry but will also lead to the creation of global recognition of the food products. Another aspect researchers are focusing on is of achieving fewer food recalls by enhancing the quality and safety of a product. This can be done by identifying its point of origin and identifying the quality of the material or the conversion of the material throughout the supply chain.

6.5.3 Lack of Communication

A major issue found in the food supply chain is fragmented information and lack of communication. Most of the stakeholders in the food supply chain have little or no information about other stakeholders' whereabouts [28]. This affects efficiency and waste generation, which can contribute to the creation of distrust among stakeholders. If the supply chain is operating globally, this problem can be increased manifold.

Technology these days provides a solution for the lack of communication. It has made it easier and faster to get an insight into the entire FSC. Live chat, microblogging and cloud-based networks are the solutions that can provide quick onboarding and an improved view of the entire supply chain, leading to ease of communication among suppliers, manufactures, distributors, retailers and the logistics partner. This, in turn, will enable smooth flow in the entire supply chain [15].

Since maintaining food quality is of utmost importance and starts with the selection of the right supplier supplying the right material, the supplier's role is very important. Food quality is closely related to the quality of agri-produce or that of the ingredients being used in food processing [20]. Therefore, to enhance quality assurance, one must get fresh ingredients of better quality from ethical suppliers. This will not only specify and maintain the quality of the raw material but will also contribute to reducing supply shortages.

6.5.4 Increase in Supply Chain Costs

There are many types of costs associated with the food supply chain: those pertaining to power, transportation, freight, human resources and adoption of the latest technology [64]. These costs affect the profitability of the supply chain, and so, keeping them under check is crucial.

When it comes to controlling costs, *knowing* the costs is the first step. Once the supply chain manager can identify the costs, managing them gets easier. Spreadsheets may be used for collecting information and managing the SC costs, but only if and when supply chains are simpler and lesser complex. As the complexity of the supply chain increases, the manager needs a technological solution. Complex supply chains are provided with a network solution that can be integrated with the supply chain. A study conducted by Chainlink Research in 2019 proposed a supply chain network termed as the network enterprise. This is a new business model that relies on the supply chain trading partner operating model, which is compatible with interoperability and visibility with reduced complexity [64]. Further, the author says that "a supply chain network needs the integration/visibility capabilities and multi-enterprise supply chain applications on one process-wide platform, a supply chain application network". The author has discussed integrator networks (IN) and "real-time single versions of the truth" networks (RSN).

Technology adoption is related to value-cost calculation. Saving too much cost can affect efficiency and growth. Adopting a technology can be an expensive affair, but it can also save costs in the long run. Also, it can modernise the business and create vulnerability to competitive forces. But, instead of focusing on cost alone, a value-cost calculation is always better for the long run.

6.5.5 Inventory Control

Inventory control is a crucial area of FSC. The leading reason that explains its importance is the perishable nature of some of the food products. Inventory needs to be managed carefully, as the supply chain manager needs to focus on controlling costs, maintaining quality and satisfying customer demands [38]. Excess inventory, beyond customer demand, leads to food spoilage and ultimately, increases food wastage. Reducing inventory beyond a certain level creates customer dissatisfaction if the customer's demands remain unfulfilled. FSC managers have a critical responsibility to manage and create a balance among customer satisfaction, inventory levels and reduction of wastage.

Ideally, to maintain and manage inventory levels in the food supply chain, there should be a system that enables and empowers supply chain visibility to the inventory through the entire supply chain using technologies like RFID, Internet of Things (IoT) and related tracking and tracing technologies that can help in maintaining and creating data accuracy. For a food supply chain manager, a challenge exists with respect to food quality and food safety for all the finished products, whether the operations are at the global level or a local level. To create effective communication throughout the supply chain, one needs to develop supply chain visibility, which will ultimately help in managing the supply chain. Creating and maintaining supplier relationships, identifying and appointing reputable and experienced logistics partners and appropriate technology adoption can lead to an efficient supply chain with improved food quality products and increased customer satisfaction. Also, it can deliver better results [26].

6.6 Conclusion

Due to basic characteristics like perishability, increased customer expectations and awareness and knowledge, the food supply chain is becoming very complex. Managing it is a challenge for supply chain managers and officials. Some of the crucial concerns are supply chain visibility, inventory control, full truckload shipping, tracking and tracing as part of traceability, lack of trust, increased intermediation, product origin provenance, food recall, food quality and food safety, all of which can increase the complexity of the SC. These complexities can be addressed and simplified with the help of technology solutions like big data, blockchain, IoT,

RFID, sensor technology, etc., apart from a few abstract solutions. Managers need to focus on technology and their relationships with customers, suppliers and logistics partners in order to enhance the efficiency, quality and safety of the food supply chain.

References

1. Ahsan K, Gunawan I (2014) Analysis of product recalls: identification of recall initiators and causes of recall. Oper Supply Chain Manag Int J 97–106. http://doi.org/10.31387/oscm0180115
2. Arduino G, Carrillo Murillo D, Parola F (2015) Refrigerated container versus bulk: evidence from the banana cold chain. Marit Policy Manag 42(3):228–245. https://doi.org/10.1080/03088839.2013.851421
3. Aung MM, Chang YS (2014) Traceability in a food supply chain: safety and quality perspectives. Food Control 39(1):172–184. http://doi.org/10.1016/j.foodcont.2013.11.007
4. Aung MM, Chang YS (2014) Temperature management for the quality assurance of a perishable food supply chain. Food Control 40(1):198–207. https://doi.org/10.1016/j.foodcont.2013.11.016
5. Badia-Melis R, Mishra P, Ruiz-García L (2015) Food traceability: new trends and recent advances. A review. Food Control 57:393–401. https://doi.org/10.1016/j.foodcont.2015.05.005
6. Barge P, Gay P, Merlino V, Tortia C (2014) Item-level radio-frequency identification for the traceability of food products: application on a dairy product. J Food Eng 125(1):119–130. https://doi.org/10.1016/j.jfoodeng.2013.10.019
7. Bennet G (2009) Food identity preservation and traceability: safer grains. https://books.google.co.in/books?hl=en&lr=&id=wREGddur_8IC&oi=fnd&pg=PP1&dq=%5B35%5D%09Bennet,+G.+S.+(2009).+Food+identity+preservation+and+traceability:+Safer+grains.+CRC+Press.&ots=FZKBcWTCLb&sig=J8syEY6zX4yUMYYaquYctkgMQ68
8. Bosona TG, Gebresenbet G (2011) Cluster building and logistics network integration of local food supply chain. Biosys Eng 108(4):293–302. https://doi.org/10.1016/j.biosystemseng.2011.01.001
9. Bosona T, Gebresenbet G (2013) Food traceability as an integral part of logistics management in food and agricultural supply chain. Food Control 33(1):32–48. http://doi.org/10.1016/j.foodcont.2013.02.004
10. Carlson CC, Peake WO (2012) Rethinking food recall communications for consumers. Iridescent 2(3):11–23. https://doi.org/10.1080/19235003.2012.11428511
11. Chammem N, Issaoui M, De Almeida AID, Delgado AM (2018) Food crises and food safety incidents in European Union, United States, and Maghreb area: current risk communication strategies and new approaches. J AOAC Int 101(4):923–938. https://doi.org/10.5740/jaoacint.17-0446
12. Charlebois S (2011) Food recalls, systemic causal factors and managerial implications: the case of premiere quality foods. Br Food J 113(5):625–636. https://doi.org/10.1108/00070701111131737
13. Chen C, Zhang J, Delaurentis T (2014) Quality control in food supply chain management: an analytical model and case study of the adulterated milk incident in China. Int J Prod Econ 152:188–199. https://doi.org/10.1016/j.ijpe.2013.12.016
14. Comba L, Belforte G, Dabbene F, Gay P (2013) Methods for traceability in food production processes involving bulk products. Biosys Eng 116(1):51–63. https://doi.org/10.1016/j.biosystemseng.2013.06.006
15. Complexities of supply chain in India and its impact on food services industry | Elior India (n.d.). Retrieved Feb 26, 2021, from https://www.elior.in/press/challenges-indian-supply-chain-and-its-impact-food-services-industry-elior-india

16. Cserháti T, Forgács E, Deyl Z, Miksik I (2005) Chromatography in authenticity and traceability tests of vegetable oils and dairy products: a review. Biomed Chromatogr 19(3):183–190. https://doi.org/10.1002/bmc.486
17. Curtin J, Kauffman RJ, Riggins FJ (2007) Making the "MOST" out of RFID technology: a research agenda for the study of the adoption, usage and impact of RFID. Inf Technol Manage 8(2):87–110. https://doi.org/10.1007/s10799-007-0010-1
18. Dabbene F, Gay P, Tortia C (2014) Traceability issues in food supply chain management: a review. Biosyst Eng 120:65–80. http://doi.org/10.1016/j.biosystemseng.2013.09.006
19. Dandage K, Badia-Melis R, Ruiz-García L (2017) Indian perspective in food traceability: a review. Food Control 71:217–227. http://doi.org/10.1016/j.foodcont.2016.07.005
20. Dharni K, Sharma RK (2015) Supply chain management in food processing sector: experience from India. Int J Logistics Syst Manag 21(1):115–132. https://doi.org/10.1504/IJLSM.2015.069080
21. Dodd MC, Bouwer JJ (2014) The supply value chain of fresh produce from field to home. Refrigeration and other supporting technologies. In: Postharvest handling: a systems approach. Elsevier Inc., pp 449–483. http://doi.org/10.1016/B978-0-12-408137-6.00016-8
22. Ellis DI, Muhamadali H, Haughey SA, Elliott CT, Goodacre R (2015) Point-and-shoot: rapid quantitative detection methods for on-site food fraud analysis-moving out of the laboratory and into the food supply chain. Anal Methods 7(22):9401–9414. http://doi.org/10.1039/c5ay02048d
23. Fitzgerald WB, Howitt OJA, Smith IJ, Hume A (2011) Energy use of integral refrigerated containers in maritime transportation. Energy Policy 39(4):1885–1896. https://doi.org/10.1016/j.enpol.2010.12.015
24. Five challenges in food and beverage supply chains (n.d.). Redwood Logistics. Retrieved Feb 26, 2021, from https://www.redwoodlogistics.com/five-challenges-in-food-and-beverage-supply-chains/
25. Fixing the 5 big problems in the food supply chain (n.d.). Retrieved Feb 26, 2021, from https://supplychainbeyond.com/5-big-problems-in-the-food-supply-chain/
26. Folkerts H, Koehorst H (1997) Challenges in international food supply chains: vertical co-ordination in the European agribusiness and food industries. Supply Chain Manag 2(1):11–14. https://doi.org/10.1108/13598549710156312
27. Food supply chain—optimizing a important supply chain (n.d.). Retrieved Feb 26, 2021, from https://www.thebalancesmb.com/food-supply-chain-2221100
28. Food supply chain disruption: challenges and learning for the F&B sector (n.d.). Retrieved Feb 26, 2021, from https://www.entrepreneur.com/article/349140
29. Fray Da Silva R, Praça I, Tsugonobu H, Yoshizaki Y, Cugnasca CE (2015) Proposal of a traceability model for the raw Brazilian sugar supply chain using RFID and WSN. In: Production and Operations Management Society, POMS 26th annual conference, Washington DC, USA
30. Frohberg K, Grote U, Winter E (2006) EU food safety standards, traceability and other regulations: a growing trade barrier to developing countries' exports? https://ageconsearch.umn.edu/record/25668/files/ip06fr01.pdf
31. Galimberti A, De Mattia F, Losa A, Bruni I, Federici S, Casiraghi M, Martellos S, Labra M (2013) DNA barcoding as a new tool for food traceability. Food Res Int 50(1):55–63. http://doi.org/10.1016/j.foodres.2012.09.036
32. Golan E, Krissoff B, Kuchler F, Nelson K, Price G, Calvin L (2003) Traceability in the US food supply: dead end or superhighway. Choices 18(2):17–20
33. Harmin C, Allen T (2015) Flood vulnerability of hog farms in Eastern North Carolina: an inconvenient poop. East Carolina University. https://thescholarship.ecu.edu/handle/10342/5143
34. James SJ, James C (2010) The food cold-chain and climate change. Food Res Int 43:1944–1956. https://doi.org/10.1016/j.foodres.2010.02.001
35. James SJ, James C (2010) Advances in the cold chain to improve food safety, food quality and the food supply chain. In: Delivering performance in food supply chains. Elsevier Ltd., pp 366–386. http://doi.org/10.1533/9781845697778.5.366

36. Kok E, Van Der Spiegel M, Prins T, Manti V, Groot M, Bremer M, Van Raamsdonk L, Van Der Fels I, Van Ruth S (2012) Traceability. In: Chemical analysis of food: techniques and applications. Elsevier Inc., pp 465–498. http://doi.org/10.1016/B978-0-12-384862-8.00014-5
37. Kozak GK, Crichton J, Farber J (2014) Control of pathogens at retail. In: Retail food safety. Springer, New York, pp 3–15. http://doi.org/10.1007/978-1-4939-1550-7_2
38. Lee HL, Billington C (1992) Managing supply chain inventory: pitfalls and opportunities. http://allman.rhon.itam.mx/~gigola/Curso_pron_inv/Pitfulls.pdf
39. Li L (2013) Technology designed to combat fakes in the global supply chain. Bus Horiz 56(2):167–177. https://doi.org/10.1016/j.bushor.2012.11.010
40. Liang K, Thomasson JA, Shen MX, Armstrong PR, Ge Y, Lee KM, Herrman TJ (2013) Ruggedness of 2D code printed on grain tracers for implementing a prospective grain traceability system to the bulk grain delivery system. Food Control 33(2):359–365. https://doi.org/10.1016/j.foodcont.2013.03.029
41. Liang K, Thomasson JA, Lee KM, Shen M, Ge Y, Herrman TJ (2012) Printing data matrix code on food-grade tracers for grain traceability. Biosys Eng 113(4):395–401. https://doi.org/10.1016/j.biosystemseng.2012.09.012
42. Lindner B, Burton M, James S, Pluske J (2001) Welfare effects of identity preservation and labelling of genetically modified food. In: 45th annual AARES conference. http://doi.org/10.22004/AG.ECON.125745
43. Maknoon MY, Soumis F, Baptiste P (2017) An integer programming approach to scheduling the transshipment of products at cross-docks in less-than-truckload industries. Comput Oper Res 82:167–179. https://doi.org/10.1016/j.cor.2017.01.018
44. Manning L (2017) Guest editorial. Br Food J 119(1):2–6. http://doi.org/10.1108/BFJ-09-2016-0446
45. Manzini R, Accorsi R (2013) The new conceptual framework for food supply chain assessment. J Food Eng 115(2):251–263. https://doi.org/10.1016/j.jfoodeng.2012.10.026
46. Miraglia M, Berdal KG, Brera C, Corbisier P, Holst-Jensen A, Kok EJ, Marvin HJP, Schimmel H, Rentsch J, Van Rie JPPF, Zagon J (2004) Detection and traceability of genetically modified organisms in the food production chain. Food Chem Toxicol 42(7):1157–1180. https://doi.org/10.1016/j.fct.2004.02.018
47. Oliver RK, Webber MD (1982) Supply-chain management: logistics catches up with strategy. In: The roots of logistics. Springer, Berlin, pp 183–194. http://doi.org/10.1007/978-3-642-27922-5_15
48. Olsen P, Borit M (2013) How to define traceability. Trends Food Sci Technol 29(2):142–150. http://doi.org/10.1016/j.tifs.2012.10.003
49. Opara LU, Mazaud F (2001) Food traceability from field to plate. Outlook Agric 30(4):239–247. https://doi.org/10.5367/000000001101293724
50. Regattieri A, Gamberi M, Manzini R (2007) Traceability of food products: general framework and experimental evidence. J Food Eng 81(2):347–356. https://doi.org/10.1016/j.jfoodeng.2006.10.032
51. Savov AV, Kouzmanov GB (2009) Food quality and safety standards at a glance. Biotechnol Biotechnol Equip 23(4):1462–1468. http://doi.org/10.2478/V10133-009-0012-8
52. Sertkaya A (2010) Food transportation safety: characterizing risks and controls by use of expert opinion. https://www.researchgate.net/publication/268179950
53. Singh P, Singh J, Antle J, Topper E, Grewal G (2014) Load securement and packaging methods to reduce risk of damage and personal injury for cargo freight in truck, container and intermodal shipments. J Appl Packag Res 6(1):47–62. http://doi.org/10.14448/japr.01.0005
54. Skoglund T, Dejmek P (2007) Fuzzy traceability: a process simulation derived extension of the traceability concept in continuous food processing. Food Bioprod Process 85(4 C):354–359. http://doi.org/10.1205/fbp07044
55. Smyth SJ, Phillips PWB (2002) Product differentiation alternatives: identity preservation, segregation, and traceability. http://www.agbioforum.missouri.edu/v5n2/v5n2a01-smyth.htm. AgBioForum. https://mospace.umsystem.edu/xmlui/handle/10355/306

56. Spink J, Moyer DC, Whelan P (2016) The role of the public private partnership in food fraud prevention—includes implementing the strategy. Curr Opin Food Sci 10:68–75. http://doi.org/10.1016/j.cofs.2016.10.002
57. Sun C, Li W, Zhou C, Li M, Yang X (2013) Anti-counterfeit system for agricultural product origin labeling based on GPS data and encrypted Chinese-sensible code. Comput Electron Agric 92:82–91. https://doi.org/10.1016/j.compag.2012.12.014
58. Supply chain management & logistics 2.0 | (2008)
59. Thakur M, Wang L, Hurburgh CR (2010) A multi-objective optimization approach to balancing cost and traceability in bulk grain handling. J Food Eng 101(2):193–200. https://doi.org/10.1016/j.jfoodeng.2010.07.001
60. The global food supply chain faces daunting challenges in the coming decades (n.d.). Retrieved Feb 26, 2021, from https://www.forbes.com/sites/stevebanker/2019/10/10/the-global-food-supply-chain-faces-daunting-challenges-in-the-coming-decades/?sh=cdbec855acf9
61. Tilman D, Cassman KG, Matson PA, Naylor R, Polasky S (2002) Agricultural sustainability and intensive production practices. Nature 418(6898):671–677. http://doi.org/10.1038/nature01014
62. Ting SL, Tse YK, Ho GTS, Chung SH, Pang G (2014) Mining logistics data to assure the quality in a sustainable food supply chain: a case in the red wine industry. Int J Prod Econ 152:200–209. https://doi.org/10.1016/j.ijpe.2013.12.010
63. Tsolakis NK, Keramydas CA, Toka AK, Aidonis DA, Iakovou ET (2014) Agrifood supply chain management: a comprehensive hierarchical decision-making framework and a critical taxonomy. Biosyst Eng 120:47–64. http://doi.org/10.1016/j.biosystemseng.2013.10.014
64. Understanding supply chain network technology—the network effect (n.d.). Retrieved Feb 26, 2021, from https://supplychainbeyond.com/understanding-supply-chain-network-technology/
65. USDA kills 73,500 Tyson-bound chickens because of bird flu | Food Safety News (n.d.). Retrieved Feb 26, 2021, from https://www.foodsafetynews.com/2017/03/usda-kills-73500-tyson-bound-chickens-because-of-bird-flu/
66. Varacca A, Soregaroli C (2016) Identity preservation in international feed supply chains. EuroChoices 15(1):38–43. https://doi.org/10.1111/1746-692X.12118
67. Viswanadham N (2006) Ed, Achieving rural and global supply chain excellence: the Indian way, Indian school of Business, December 5, 2006 (released by the Prime Minister of India). https://www.csa.iisc.ac.in/~nv/Mypublications/Books/GLAMSBook.pdf
68. Walaszczyk A, Galińska B (2020) Food origin traceability from a consumer's perspective. Sustainability 12(5):1872. https://doi.org/10.3390/su12051872
69. Wang J, Yue H, Zhou Z (2017) An improved traceability system for food quality assurance and evaluation based on fuzzy classification and neural network. Food Control 79:363–370. https://doi.org/10.1016/j.foodcont.2017.04.013
70. Yaacob TZ, Suzana Jaafar H, Rahman FA (2016) An overview of halal food product contamination risks during transportation. Sci Int (Lahore) 28(3):3183–3190
71. Ye Y, Li J, Li K, Fu H (2018) Cross-docking truck scheduling with product unloading/loading constraints based on an improved particle swarm optimisation algorithm. Int J Prod Res 56(16):5365–5385. https://doi.org/10.1080/00207543.2018.1464678
72. Zhou GH, Xu XL, Liu Y (2010) Preservation technologies for fresh meat—a review. Meat Sci 86(1):119–128. http://doi.org/10.1016/j.meatsci.2010.04.033
73. Zulfakar MH, Anuar MM, Talib MSA (2014) Conceptual framework on halal food supply chain integrity enhancement. Procedia Soc Behav Sci 121:58–67. https://doi.org/10.1016/j.sbspro.2014.01.1108

Chapter 7
A Framework for Blockchain Technology Adoption in Food Supply Chain

Ashish Dwivedi, Ajay Jha, Vikas Yadav, Manisha Bansal, and Harsh Srivastava

Abstract With its several unique and useful applications, blockchain has proven to be a positively disruptive technology in various sectors. Blockchain applications for the food sector had been successfully developed and practised numerous times. The stakeholders of the food supply chain (FSC) are actively exploring the diffusion strategy for the blockchain. The present chapter identifies and analyses the Critical Success Factors (CSFs) that can facilitate the successful adoption of Block Chain Technology (BCT) in FSC. The results of the study highlight that the "Technological maturity (CSF3)" and "Resiliency (CSF5)" are the potential driving CSFs while "Information security and reliability (CSF1)" and "Trust and collaboration (CSF9)" are the most dependent CSFs for BCT adoption in FSC. The analysis from the study draw the contextual and hierarchical relationships among the identified CSFs of BCT adoption in FSC. The outcomes can assist the policymakers in the successful dissemination of BCT and enhancing the performance of FSC.

Keywords Food supply chain · Blockchain · Interpretive structural modelling · Information system · Drivers · Theory building

A. Dwivedi (✉)
Jindal Global Business School, O. P. Jindal Global University, Sonipat, India

A. Jha
Operations Management, Jaipuria Institute of Management, Lucknow, India

V. Yadav
Indian Railway, New Delhi, India

M. Bansal
Indian Institute of Technology Delhi, New Delhi, India
e-mail: manisha.bansal@alumni.iitd.ac.in

H. Srivastava
Organization Design, GMR Group, New Delhi, India

R. S Mor et al. (eds.), *Operations and Supply Chain Management in the Food Industry*, Lecture Notes in Management and Industrial Engineering,
https://doi.org/10.1007/978-981-16-5555-5_7

7.1 Introduction

In the present scenario, food supply chain (FSC) industry merely utilizes advances in Information and Communication Technology (ICT) to hoard and track the order deliveries. Despite having abundant technology to support productive features such as tracking, provenance, automatic audit, automatic transaction, quality monitoring and smart contract execution, FSCs are yet to facilitate these features for its stakeholders [10]. These features are supported by combinations of blockchain technology (BCT) with the Internet of Things (IoT) and artificial intelligence (AI). The application of BCT and AI can substantially improve the performance of FSC [1]. Several authors anticipated the increase in demand for applications based on BCT and AI technologies [23, 49, 53]. Therefore, researchers around the globe have extended their attention towards the development of innovative applications, best practices and procedures, theoretical foundations, practical considerations and others aspects for successful deployment of these technologies [4, 22, 41, 48].

BCT is anticipated to bring substantial improvement in terms of accountability, transparency, trust and resiliency of FSC. Further, BCT can bring the information power symmetry along the supply chain irrespective of the size and scope of the stakeholders [8, 9, 23]. The researchers have explored the BCT application in food quality monitoring, land and use record, provenance tracking, financial transactions, pesticide and farm use control, farm to fork tracking, etc. [10, 32]. In recent years, the previous literature has worked on the prospects, benefits and complexities of the BCT deployment [3, 37, 50]. However, the theoretical foundation for the successful dissemination of BCT along the FSC remains an overlooked aspect. There are limited studies that had addressed the enablers, CSFs and decision-making frameworks for the adoption of BCT [23]. Therefore, it is essential to have an understanding of the CSFs for resolving the issues in context of FSC. Further, understanding the dynamics among the CSFs of BCT adoption, their interrelationships, scope and influences, can assist the BCT developers in convincing the FSC stakeholders to implement the BCT. Therefore, the present chapter aims to attempt the following objectives:

1. What are the Critical Success Factors (CSFs) of BCT implementation in FSC?
2. How useful are these CSF and their interactions for the practitioners and decisionmakers?

7.2 Literature Review

The previous literature observes significant contributions related to blockchain adoption in different domains. The researchers have worked on the application development, theoretical discussions, pilot studies, blueprints, simulation and modelling approaches [29–31, 41]. There are several studies evident in literature in context of food and agriculture sector. However, there is a dearth of literature assessing the CSFs affecting blockchain adoption in the context of FSC. Kamble et al. [23] modelled

Table 7.1 CSFs of BCT adoption in FSC

S. No.	Critical Success Factors (CSFs)	References
1	Information security and reliability (CSF1)	[17]
2	Faster transaction processing system (CSF2)	[18, 47]
3	Technological maturity (CSF3)	[11, 48, 56]
4	Provenance recording and tracking (CSF4)	[17, 21, 39]
5	Resiliency (CSF5)	[28, 38]
6	Quality and safety management (CSF6)	[20, 48, 55]
7	Interoperability (CSF7)	[2, 26]
8	Smart contract (CSF8)	[19, 27]
9	Trust and collaboration (CSF9)	[6, 7, 13]
10	Ease of documentation (CSF10)	[51]
11	Immutability (CSF11)	[10, 28]
12	Intermediary reduction (CSF12)	[27, 44, 50]

Source Author

the enablers of BCT in the context of agriculture supply chain. However, there is a limited scope to BCT-enabled traceability for attaining sustainability in the supply chain. A comprehensive study accessing the attributes and interrelationship among the CSFs affecting the BCT adoption is still lacking.

7.2.1 *Critical Success Factors for BCT Adoption*

A detailed literature survey was conducted for the identification of CSFs for BCT adoption in FSC. The list of CSFs identified from the literature review is presented in Table 7.1. The identified CSFs drive the adoption of BCT by the organizations engaged in the FSC.

A detailed discussion on the scope and characteristics of the identified CSFs for BCT adoption in FSC is presented below:

7.2.1.1 Information Security and Reliability (CSF1)

The adulteration of food and fraud along the supply chain has become a major cause of concern for the FSC stakeholders [16, 17]. BCT key features to provide information in a secure, transparent and reliable manner that can positively affect the functioning of the FSC. Therefore, it can be included as a potential CSF for BCT adoption.

7.2.1.2 Faster Transaction Processing System (CSF2)

FSC has become more complex on account of diversified food demand around the globe. The global flow of food has generated the requirement to have tight control over the maintenance of food quality, transparency and faster transaction speed [47]. Therefore, BCT can serve as a critical determinant of operational performance by easing the custom process and removing the international transaction complexities [18].

7.2.1.3 Technological Maturity (CSF3)

BCT observes several constraints that can be attributed to the nascent stage of technological development. Few critical issues that limit the deployments are power hungry nature of technology, intensive computing power requirement and infrastructure requirements [11, 48, 56]. Also, the wide power gap among the FSC stakeholders and derived control over the network can cause several issues with BCT.

7.2.1.4 Provenance Recording and Tracking (CSF4)

Increased awareness regarding the ethical attributes of food and concerns related to hygiene and food adulteration can be resolved by the deployment of BCT [23, 54]. In recent years, several pilot studies are successfully conducted to facilitate food provenance and tracking [17, 21, 39]. Therefore, provenance recording and tracking qualify as a CSF for BCT adoption in FSC.

7.2.1.5 Resiliency (CSF5)

BCT offers higher resiliency than the legacy system currently installed in FSC [28]. Improved resiliency of transaction settlement enables better control over the quality of food and satisfaction of customers, thus facilitating performance improvement of FSC and better response to uncertainty and disruptions [38].

7.2.1.6 Quality and Safety Management (CSF6)

Food quality, cleanliness and authenticity are a competitive advantage in the global FSC [20]. Several technological solutions contribute to the safety and quality management of food. Radio Frequency Information Device (RFID), bar code, IoT and others serve to collect the real-time data and other tracing and tractability systems. Blockchain holds the potential to empower these systems with more transparency, security and authenticity of data, thus enabling an improved quality and safety management system [48, 55].

7.2.1.7 Interoperability (CSF7)

The globalization of FSC has created information asymmetry between producer, service provider and consumer [26]. The FSC needs to coordinate asymmetric stakeholders, transporters, government and regulatory body and diversified consumer. Thus, there is a requirement for interoperability standards to expedite business-to-business and business-to-consumer integration in FSC. BCT might enable the FSC stakeholders to evolve with the ever-changing FSC. BCT application can serve as a transparent, robust and efficient information system [2].

7.2.1.8 Smart Contract (CSF8)

BCT increases the value of certification and transaction settlement times through smart contracts [19]. Smart contract feature facilitates the automatic business process. Also, smart contract feature can report mitigating the food fraud and unethical business process [27]. The developers are performing experiments by supporting the small-scale farmers with financial grant and later integrating them in a productive and economically beneficial ecosystem. Therefore, it is apparent that smart contract feature can transform the FSC and benefit several stakeholders [27].

7.2.1.9 Trust and Collaboration (CSF9)

The majority of legacy systems adopted in FSC heavily rely on the centralized information system that suffers from lack of trust, security threats, confidentiality and auditability [10]. BCT intrinsic capability to increase trust and collaboration among supply chain partners presents an unavoidable scenario for their adoption in the supply chain [6, 7, 13].

7.2.1.10 Ease of Documentation (CSF10)

The supply chain gains efficiency by digitalizing documents transfers and accelerating data in a globally complex supply chain [51]. The reduction of trade documents and processing cost eliminates the delays associated with paper movement in international transactions, thus enabling better visibility and transparency. This results in ease of documentation as a potential CSF for FSC.

7.2.1.11 Immutability (CSF11)

BCT by design is immutable and supports easy auditability. This transparent source of information resolves the issues of power and information asymmetry along the

Table 7.2 Details of experts and their designation

S. No.	Designation	Work experience (Years)
1	Government Mandi officer	12
2	Transporter	20
3	Blockchain developer	5
4	Research scholar	3
5	Traders	8

Source Author

FSC stakeholders [10], thus facilitates the better coordination, reliability and trust in the network [28].

7.2.1.12 Intermediary Reduction (CSF12)

In a few instances, intermediary reported to follow the unethical business process and generated inefficiencies along the supply chain [51]. The power and money acquisition by intermediary sometimes results in exploitation of low and small producers [44]. Also, these intermediaries cause a loss in quality of food. Also, the reduction of intermediary by adopting BCT can improve the performance and benefits of several stakeholders of FSC [27].

7.3 Data Collection and Sampling Design

The data for the present study is gathered by arranging scheduled interviews with closed-ended questionnaires. A closed-ended questionnaire was developed to ascertain the effect of all CSFs concerning each other, which are said to be pairwise comparisons. In total, five experts are selected for further analysis. The details of the experts' are presented in Table 7.2.

The experts were selected based on their extensive experience in FSC. Blockchain developer and research scholar made the expert panel multidimensional. They provided several insights on the BCT application and how BCT adoption affects the FSC. Approximately 45–70 min were spent for each expert. A total of 300 min were spent on interviewing the experts.

7.4 Methodology

This section briefly explains the methodology adopted in this chapter. The targeted objectives of this chapter are achieved by employing a modified version of Interpretive Structure Modelling (ISM). ISM is a widely used method to construct the hierarchical framework among factors/drivers/barriers. ISM method also produces the interrelationship among the factors. ISM is a constructive approach when dealing with convoluted problems but inefficient in constructing a theory to answer why. Consequently, it is necessary to abstractly represent the links to define the appropriate interactions among the elements as suggested by the experts [45].

In order to overcome the shortcomings of ISM technique an extended form "Total Interpretive Structural Modelling (TISM)" is introduced. In the TISM technique, the ISM limitation is focused by using the "Interpretive Matrix" tool where causal thinking is confined at the time of data collection by experts and made explicit in the matrix [45]. Some recent publications adopted the TISM methodology for solving complicated problems. Chaple et al. [12] modelled the lean barriers for successful lean implementation. Bag [5] obtained the interactions among the green procurement drivers using the TISM methodology. Workforce agility IoT project enablers were modelled adopting TISM method [35]. A sustainable manufacturing evaluation model was obtained for leather industries adopting TISM technique [14, 15]. Interactions among the antecedents of green human resource management were established employing the TISM technique [33].

Further, Sushil [46] modified TISM and added simultaneous transitive check. This modified version reduced the need of pairwise comparison. This reduction in pairwise comparisons results in the reduction of errors with human judgement, thus making the analysis more efficient and transparent. Several authors have recently adopted mTISM methodology. For more details on mTISM, readers are advised to refer [36, 42].

Next, we describe stepwise result of mTISM exercise implemented in this chapter.

Step I: Identify and define CSF

The first step of mTISM is to identify and define the key factors of the objectives. A total of twelve CSFs are identified from the study related to blockchain adoption in FSC from personal interviews, and literature review as shown in Table 7.1.

Step II: Pairwise comparison

A relationship among each pair of CSFs reflecting an existence or deficiency of relationship is performed. This step involves the simultaneous transitive check. In this modified form, there is reduction of pairwise comparison of the pairs that have transitive links. This assists in construction of reachability matrix. The final reachability matrix obtained for this study is shown in Table 7.3.

Step III: Partitioning the reachability matrix

This step consists of making reachability set, antecedent set and intersection set. The reachability matrix constructed in the previous step is partitioned into different levels using number of iterations based on the following principle. Table 7.4 represents the levels of CSF obtained during this exercise.

Table 7.3 Final reachability matrix for CSFs

	Information security and reliability (CSF1)	Faster transaction processing system (CSF2)	Technological maturity (CSF3)	Provenance recording and tracking (CSF4)	Resiliency (CSF5)	Quality and safety management (CSF6)	Interoperability (CSF7)	Smart contract (CSF8)	Trust and collaboration (CSF9)	Ease of documentation (CSF10)	Immutability (CSF11)	Intermediary reduction (CSF12)
Information security and reliability (CSF1)	1	0	0	0	1	0	0	0	1*	0	0	0
Faster transaction processing system (CSF2)	1	1	0	1	1*	1*	0	1*	1*	1*	0	1*
Technological maturity (CSF3)	1*	1	1	1	1*	1	1	1*	1*	1*	0	0
Provenance recording and tracking (CSF4)	1*	1	0	1	1	1	0	1	1*	1*	0	1
Resiliency (CSF5)	1	0	0	0	1	0	0	0	1	0	0	0
Quality and safety management (CSF6)	1*	0	0	0	1	1	0	0	1	0	0	0
Interoperability (CSF7)	1	1	0	1	0	0	1	1	1*	1*	0	1
Smart contract (CSF8)	1*	0	0	0	1	1	0	1	1	1	0	0

(continued)

Table 7.3 (continued)

	Information security and reliability (CSF1)	Faster transaction processing system (CSF2)	Technological maturity (CSF3)	Provenance recording and tracking (CSF4)	Resiliency (CSF5)	Quality and safety management (CSF6)	Interoperability (CSF7)	Smart contract (CSF8)	Trust and collaboration (CSF9)	Ease of documentation (CSF10)	Immutability (CSF11)	Intermediary reduction (CSF12)
Trust and collaboration (CSF9)	1*	0	0	0	1	0	0	0	1	0	0	0
Ease of documentation (CSF10)	1*	0	0	0	1*	1*	0	1	1	1	0	0
Immutability (CSF11)	1*	1*	0	1	1*	1*	0	1*	1*	1	1	1
Intermediary reduction (CSF12)	1*	0	0	0	1*	1*	0	1*	1*	1	0	1

Source Author
1 represents a direct link among factors, 0 represents no link among the factors, and 1* represents a transitive link

Table 7.4 Iteration (1–6) for the CSFs

CSF	Reachability set	Antecedent set	Intersection set	Level
CSF1	1, 5, 9	1, 2, 3, 4, 5, 6, 7, 8, 9, 10, 11, 12	1, 5, 9	L1
CSF2	1, 2, 4, 5, 6, 8, 9, 10, 12	2, 3, 4, 7, 11	2, 4	L4
CSF3	1, 2, 3, 4, 5, 6, 7, 8, 9, 10	3	3	L7
CSF4	1, 2, 4, 5, 6, 8, 9, 10, 12	2, 3, 4, 7, 11	2, 4	L5
CSF5	1, 2, 3, 4, 5, 6, 8, 9, 10, 11, 12	1, 5, 9	1, 5, 9	L7
CSF6	1, 5, 6, 9	2, 3, 4, 6, 8, 10, 11, 12	6	L2
CSF7	1, 2, 4, 7, 8, 9, 10, 12	3, 7	7	L6
CSF8	1, 5, 6, 8, 9, 10	2, 3, 4, 7, 8, 10, 11, 12	8, 10	L3
CSF9	1, 5, 9	1, 2, 3, 4, 5, 6, 7, 8, 9, 10, 11, 12	1, 5, 9	L1
CSF10	1, 5, 6, 8, 9, 10	2, 3, 4, 7, 8, 10, 11, 12	8, 10	L3
CSF11	1, 2, 4, 5, 6, 8, 9, 10, 11, 12	11	11	L6
CSF12	1, 5, 6, 8, 9, 10, 12	2, 4, 7, 11, 12	12	L4

Source Author

Step IV: Development of Interpretive Structure Model

In last step, an ISM model is constructed using the relationships obtained in the last steps. This model depicts the various direct and transitive relationship among the CSFs. Figure 7.1 represents the ISM model of the CSFs for successful BCT adoption in FSC.

Table 7.5 explains the key insights obtained during the interview with experts. The experts suggested their opinion regarding the basis of relationship among the CSFs. Further, several insights were matched and tabulated for easy understanding of reader in Table 7.5. The first row and first column contain the CSFs analysed in the chapter. The corresponding block to each row and column explains the mechanism or nature of relationships.

7.5 Results and Discussion

BCT is at a nascent stage in the FSC and counters several challenges that are required for efficient implementation. Several researchers and experts have highlighted the benefits of BCT surpass the challenges and complexities of their adoption in FSC [40]. With appropriate policies and strategic support, FSC stakeholders can leverage the BCT potential to full extent. To meet that end, this chapter provides a model of CSFs that requires consideration. This chapter provides a theoretical foundation to counter the practical problems that may occur in BCT adoption. In this section, outcomes from the study, results and discussion and the insights are discussed.

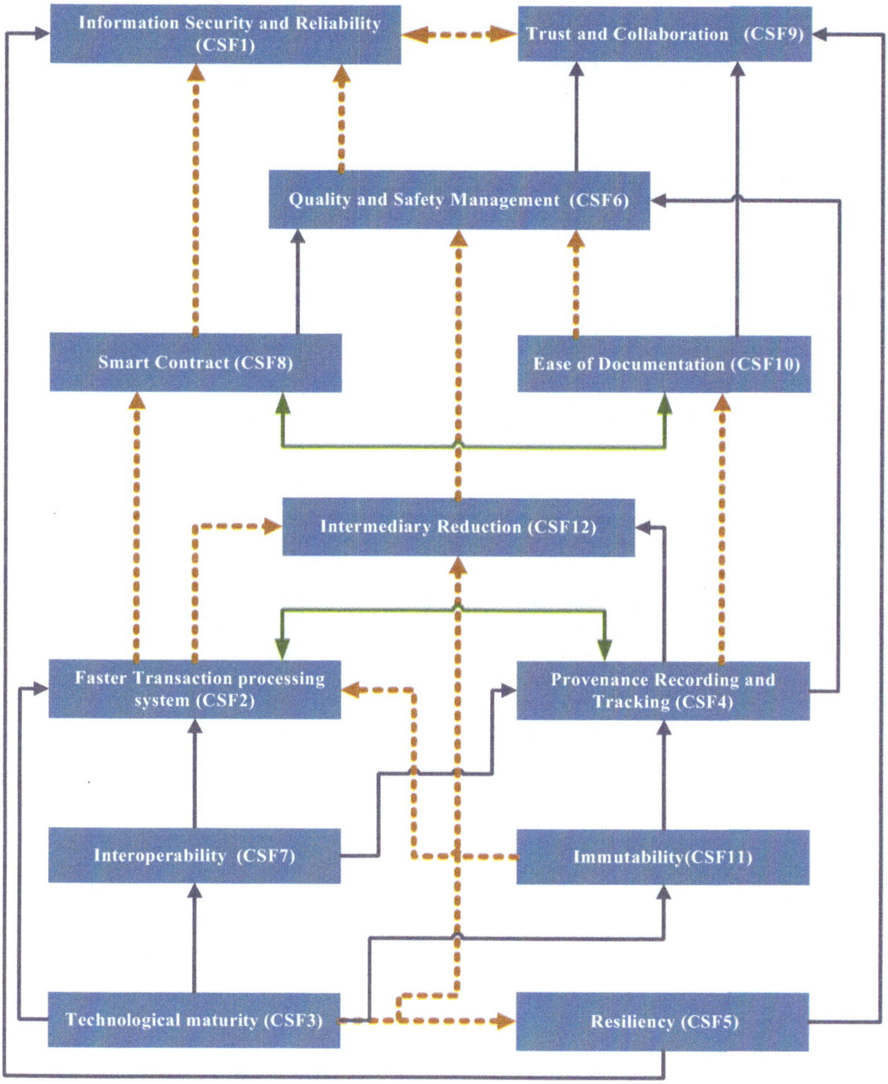

Fig. 7.1 ISM model reflecting relationships among the CSFs. *Source* Author

In this chapter, to analyse the interrelationships among different CSFs of BCT in FSC, a mTISM model has been developed. Based on the mTISM results, CSFs are arranged as follows. CSF1 > CSF9 > CSF6 > CSF8 > CSF10 > CSF12 > CSF2 > CSF4 > CSF7 > CSF11 > CSF3 > CSF5. The mTISM model can be studied in seven different levels. Information security and reliability (CSF 1) and Trust and collaboration (CSF 9) are at the first level. Its position on the topmost level signifies that this CSF is affected by rest of the CSFs studied in this chapter. This result is in

Table 7.5 Explanation of interrelationship among the CSFs

	Information security and reliability (CSF1)	Faster transaction processing system (CSF2)	Technological maturity (CSF3)	Provenance recording and tracking (CSF4)	Resiliency (CSF5)
Information security and reliability (CSF1)	1	0	0	0	Data reliability enhances the organization capabilities against malicious attacks and disruption
Faster transaction processing system (CSF2)	Faster processing reduce the risk of tempering and unauthorized access	1	0	Improved speed of information and transaction flow ease the tacking process	Transitive
Technological maturity (CSF3)	Transitive	Innovation and increased attention from researchers will reduce the transaction time. Further facilitating faster payments to parties	1	Mature technology improves the facilitations of tracking and tracking capabilities	Transitive
Provenance recording and tracking (CSF4)	Transitive	Capabilities assist the stakeholders in decision-making regarding the financial transaction	0	1	Improved tracking capabilities improves the organizational readiness for future disruptions, uncertainty and fluctuation
Resiliency (CSF5)	Resilient supply chain increases the customers' perceptions and in turn makes system more reliable	0	0	0	1

(continued)

Table 7.5 (continued)

	Information security and reliability (CSF1)	Faster transaction processing system (CSF2)	Technological maturity (CSF3)	Provenance recording and tracking (CSF4)	Resiliency (CSF5)
Quality and safety management (CSF6)	Transitive	0	0	0	Better control over quality of the products and safety of products increases the chances of responding to uncertainty appropriately
Interoperability (CSF7)	BCT capability to keep record immutable improves the system reliability	Interoperability assist in transaction speed	0	Interoperability among different systems and devices is must for tracking across wide range of users	0
Smart contract (CSF8)	Transitive	0	0	0	Removes the risk of fraudulent activities thus improving system responsiveness
Trust and collaboration (CSF9)	Transitive	0	0	0	Collaboration among stakeholders helps in countering disruptions
Ease of documentation (CSF10)	Transitive	0	0	0	Transitive
Immutability (CSF11)	Transitive	Transitive	0	Enhances the reliability of tracking facilities	Transitive
intermediary reduction (CSF12)	Transitive	0	0	0	Transitive

(continued)

Table 7.5 (continued)

	Quality and safety management (CSF6)	Interoperability (CSF7)	Smart contract (CSF8)	Trust and collaboration (CSF9)
Information security and reliability (CSF1)	0	0	0	Transitive
Faster transaction processing system (CSF2)	Transitive	0	Transitive	Transitive
Technological maturity (CSF3)	Maturity in technology enhances the quality control practices	Technological innovations with time enhance the applications integration thus improving the supply chain and adoption of BCT	Transitive	Transitive
Provenance recording and tracking (CSF4)	Improved tracking capabilities enhances the quality control practices and ensures the customer satisfaction regarding food quality	0	Tracking capabilities support the smart contracting activities	Transitive
Resiliency (CSF5)	0	0	0	Resiliency in system enhances the swift trust
Quality and safety management (CSF6)	1	0	0	Better control on quality generated the swift trust among stakeholders
Interoperability (CSF7)	0	1	To facilitate the smart contracting. Lack of interoperability reduces the utility of smart contracting feature	Transitive

(continued)

Table 7.5 (continued)

	Quality and safety management (CSF6)	Interoperability (CSF7)	Smart contract (CSF8)	Trust and collaboration (CSF9)
Smart contract (CSF8)	Reduces the risk of tempering and unauthorized transactions and unauthorized tempering with products and the delay associated with quality check and transaction	0	1	Contracting feature enhances the trust among stakeholders and improves the operations performance of SC
Trust and collaboration (CSF9)	0	0	0	1
Ease of documentation (CSF10)	Transitive	0	Improve the responsiveness of smart contracting features	Paperwork reduction, transparency and rapid flow of information enhances the collaboration among stakeholders
Immutability (CSF11)	Transitive	0	Transitive	Transitive
Intermediary reduction (CSF12)	Transitive	0	Transitive	Transitive

	Ease of documentation (CSF10)	Immutability (CSF11)	Intermediary reduction (CSF12)
Information security and reliability (CSF1)	0	0	0
Faster transaction processing system (CSF2)	Transitive	0	Transitive
Technological maturity (CSF3)	Transitive	0	0
Provenance recording and tracking (CSF4)	Transitive	0	Removes the assistance of any third party for tracking activities
Resiliency (CSF5)	0	0	0
Quality and safety management (CSF6)	0	0	0

(continued)

Table 7.5 (continued)

	Ease of documentation (CSF10)	Immutability (CSF11)	Intermediary reduction (CSF12)
Interoperability (CSF7)	Transitive	0	Integrations among key stakeholders removes the nonessential intermediary
Smart contract (CSF8)	Reduces the need of documentation	0	0
Trust and collaboration (CSF9)	0	0	0
Ease of documentation (CSF10)	1	0	0
Immutability (CSF11)	Reduces the resources enjoyment in inspection and monitoring activities	1	Removes the needs of any regulatory and controlling authority. Thus improving the operational performance
Intermediary reduction (CSF12)	Limited stakeholders improve the documentation	0	1

Source Author

line with the study conducted by Shoaib et al. [43]. Further, it defines that this CSF has least influence on other CSFs. Quality and safety management (CSF 6) is at the second level. Smart contract (CSF 8) and Ease of documentation (CSF 10) are at the third level. Intermediary reduction (CSF 12) is at the fourth level. Faster transaction processing system (CSF 2) and Provenance recording, and tracking (CSF 4) are at the fifth level. Interoperability (CSF 7) and Immutability (CSF 11) are at the sixth level. Technological maturity (CSF 3) and Resiliency (CSF 5) are at the seventh and the bottom level in the mTISM structure.

The results from the study highlight that the reliability and security of the information is the most crucial feature of the BCT that will attract the FSC stakeholders. Also, this outcome has been supported by Kamble et al. [24], Nakasumi [34], Weber et al. [52]. In literature, previous studies had raised several concerns while maintaining the food quality in transit. This becomes more difficult in the global and disconnect supply chain [25]. Thus, safety and quality of food can be a differentiating factor for organizations. mTISM analysis exposes this significance of BCT enhanced food quality which has been largely overlooked in the previous literature. The findings suggest that BCT developers are required to pay more attention to the BCT feature that enhances the food quality and safety [17]. Further, reduction of documents and execution of rapid transactions will also become a potential feature that will be sought by various stakeholders. For the FSC stakeholders and BCT developers, several insights emerged during the semi-structured interview process. One respondent suggested that the sensitization of regulatory and intermediary body is essential to meet the desired expectations from the BCT. Other supporting infrastructure needs to be built around the whole supply chain network. This requires the development of robust IT network and innovations in IoT devices considering the FSC requirements. Government also needs to be proactive for its policies, regulations and infrastructure development support. Also, the regulations need to be modified that prevent or limit the BCT-based transaction. The same has been suggested by Rugeviciute and Mehrpouya [37]. Another suggestion for the blockchain developers emerges regarding considerations of small producers. BCT developers need to incorporate the features that can easily integrate small prouder and secure their economic well-being. FSC struggles with the domination of few influencing stakeholders who sometimes engage in unethical practices. With BCT deployment, influencing stakeholders can derive enormous power and might end up causing more complications.

7.6 Conclusion, Limitations and Future Research Directions

The food supply chain is critical due to involvement of public health and safety. Thus, it requires special attention for flawless operations of supply chain system. To operationalize fluently, various emerging technologies have been implemented within the system. Among them, BCT is promising to cater issues related to FSC. This chapter aims to explore, understand and model the CSFs of BCT adoption in FSC. In this chapter, a literature review was performed to identify the CSFs of BCT in FSC. Interrelationships among the CSFs are presented using the final reachability matrix (Table 7.3). Finally, mTISM model reflecting interactions among the CSF appropriate to FSC is obtained (Fig. 7.1). The chapter is unique in terms of its focus towards FSC. The provided insights will help the FSC stakeholders in the BCT adoption process.

The BCT would be helpful in various aspects to the existing supply chain. It can be used for financial services such as credit transfer using smart contract system. This will not only reduce the intermediaries but also provide secure transactions between the parties. The technology can also be used to reduce the paperwork and documentation process. This will lead to reduction in overall lead time as well as ensure the quality. Addition to this, several other technologies can also be clubbed with BCT to provide ease of use and transparency. For an instance, IoT can be used along with BCT to capture data at each stage and share it with required stakeholders. This will prevent the loss of data with secure record keeping as a transaction in blockchain system. This chapter highlights the various uses of BCT considering different factors which can make its implementation successful.

The key limitation of this study is the respondent geographical diversity. However, it helped in generating insights for a very complex FSC in India. In India, FSC is characterized with large number of small-scale producers and corrupt intermediary. These insights serve well for the scope of the study. However, future researchers are recommended to explore the CSFs from different geographical perspectives.

References

1. Ahumada O, Villalobos JR (2009) Application of planning models in the agri-food supply chain: a review. Eur J Oper Res 196(1):1–20. http://doi.org/10.1016/j.ejor.2008.02.014
2. Ande R et al (2020) Internet of things: evolution and technologies from a security perspective. Sustain Cities Soc 54:101728. http://doi.org/10.1016/j.scs.2019.101728
3. Angelis J, Ribeiro da Silva E (2018) Blockchain adoption: a value driver perspective. Bus Horiz 62(3):307–314. http://doi.org/10.1016/j.bushor.2018.12.001
4. Antonucci F et al (2019) A review on blockchain applications in the agri-food sector. J Sci Food Agric 99(14):6129–6138. https://doi.org/10.1002/jsfa.9912
5. Bag S (2019) Identification of green procurement drivers and their interrelationship using fuzzy TISM and MICMAC analysis. In: Advanced methodologies and technologies in engineering and environmental science. IGI Global, pp 167–186
6. Baharmand H, Comes T (2019) Leveraging partnerships with logistics service providers in humanitarian supply chains by blockchain-based smart contracts. IFAC-PapersOnLine 52(13):12–17. https://doi.org/10.1016/j.ifacol.2019.11.084
7. Biggs J et al (2017) Blockchain: revolutionizing the global supply chain by building trust and transparency, pp 1–26
8. Bronson K, Knezevic I (2016) Big data in food and agriculture, pp 1–5. http://doi.org/10.1177/2053951716648174
9. Carbonell IM (2016) The ethics of big data in big agriculture. Internet Policy Rev 5(1):1–13. https://doi.org/10.14763/2016.1.405
10. Caro MP et al (2018) Blockchain-based traceability in agri-food supply chain management: a practical implementation. IEEE, pp 2018–2021
11. Chamberlain SL (2019) Assessing the merits of blockchain technology for global sustainable development initiatives. Harvard Extension School. Available at: http://nrs.harvard.edu/urn-3:HUL.InstRepos:42004229%0A
12. Chaple AP et al (2018) Interpretive framework for analyzing lean implementation using ISM and IRP modeling. Benchmarking 25(9):3406–3442. https://doi.org/10.1108/BIJ-07-2017-0177
13. Dubey R et al (2020) Blockchain technology for enhancing swift-trust, collaboration and resilience within a humanitarian supply chain setting. Int J Prod Res. http://doi.org/10.1080/00207543.2020.1722860
14. Dwivedi A, Agrawal D, Madaan J (2019) Sustainable manufacturing evaluation model focusing leather industries in India: a TISM approach. J Sci Technol Policy Manag 10(2):319–359. https://doi.org/10.1108/JSTPM-06-2018-0054
15. Dwivedi A, Madaan J (2020) A hybrid approach for modeling the key performance indicators of information facilitated product recovery system. J Model Manag 15(3):933–965. https://doi.org/10.1108/JM2-01-2019-0003
16. Dwivedi A, Jha A, Prajapati D, Sreenu N, Pratap S (2020) Meta-heuristic algorithms for solving the sustainable agro-food grain supply chain network design problem. Mod Supply Chain Res Appl 2(3):161–177. https://doi.org/10.1108/MSCRA-04-2020-0007
17. Galvez JF, Mejuto JC, Simal-Gandara J (2018) Future challenges on the use of blockchain for food traceability analysis. TrAC Trends Anal Chem 107:222–232. http://doi.org/10.1016/j.trac.2018.08.011

18. Ganne E (2018) Can blockchain revolutionize trade? Available at: http://blogs.worldbank.org/trade/can-blockchain-revolutionize-trade
19. Ge L et al (2017) Blockchain for agriculture and food: findings from the pilot study. Wageningen Economic Research. Available at: www.wur.eu/economic-research
20. George RV et al (2019) Food quality traceability prototype for restaurants using blockchain and food quality data index. J Cleaner Prod 240:118021. http://doi.org/10.1016/j.jclepro.2019.118021
21. Kamath R (2018) Food traceability on blockchain: Walmart's pork and mango pilots with IBM. J Br Blockchain Assoc 1(1):1–12. https://doi.org/10.31585/jbba-1-1-(10)2018
22. Kamble SS, Gunasekaran A, Gawankar SA (2020) Achieving sustainable performance in a data-driven agriculture supply chain: a review for research and applications. Int J Prod Econ 219(July 2018):179–194. http://doi.org/10.1016/j.ijpe.2019.05.022
23. Kamble SS, Gunasekaran A, Sharma R (2019) Modeling the blockchain enabled traceability in agriculture supply chain. Int J Inf Manag 52(June 2019):101967. http://doi.org/10.1016/j.ijinfomgt.2019.05.023
24. Kamble SS, Gunasekaran A, Sharma R (2020) Modeling the blockchain enabled traceability in agriculture supply chain. Int J Inf Manag 52(May 2019):101967. http://doi.org/10.1016/j.ijinfomgt.2019.05.023
25. Kayikci Y et al (2020) Food supply chain in the era of Industry 4.0: blockchain technology implementation opportunities and impediments from the perspective of people, process, performance, and technology. Prod Plan Control 1–21. http://doi.org/10.1080/09537287.2020.1810757
26. Keogh JG et al (2020) Data and food supply chain: blockchain and GS1 standards in the food chain: a review of the possibilities and challenges. Build Future Food Saf Technol 145–178. http://doi.org/10.1016/B978-0-12-818956-6.00007-5
27. Kim HM, Laskowski M (2018) Agriculture on the blockchain: sustainable solutions for food, farmers, and financing. Supply Chain Revolution, Barrow Books
28. Kumar A (2019) Is blockchain a silver bullet for supply chain management? Technical challenges and research opportunities. Decis Sci 1–30. http://doi.org/10.1111/deci.12396
29. Kuo T, Kim H, Ohno-machado L (2017) Blockchain distributed ledger technologies for biomedical and health care applications. J Am Med Inform Assoc 24(6):1211–1220. https://doi.org/10.1093/jamia/ocx068
30. Longo F et al (2019) Blockchain-enabled supply chain: an experimental study. Comput Ind Eng 136(July):57–69. https://doi.org/10.1016/j.cie.2019.07.026
31. L'Hermitte C, Nair N-KC (2020) A blockchain-enabled framework for sharing logistics resources in emergency operations. Disasters 1–31. http://doi.org/10.1111/disa.12436
32. Maru A et al (2018) Digital and data-driven agriculture: harnessing the power of data for smallholders. F1000Research 7(525):38. Available at: https://cgspace.cgiar.org/bitstream/handle/10568/92477/GFAR-GODAN-CTA-white-paper-final.pdf%0A https://f1000research.com/documents/7-525
33. Moktadir MA et al (2019) Antecedents for greening the workforce: implications for green human resource management. Int J Manpow. https://doi.org/10.1108/IJM-07-2019-0354
34. Nakasumi M (2017) Information sharing for supply chain management based on block chain technology. In: Proceedings—2017 IEEE 19th conference on business informatics, CBI 2017, vol 1, pp 140–149. http://doi.org/10.1109/CBI.2017.56
35. Patil M, Suresh M (2019) Modelling the enablers of workforce agility in IoT projects: a TISM approach. Glob J Flex Syst Manag 20(2):157–175. https://doi.org/10.1007/s40171-019-00208-7
36. Patil A, Shardeo V, Madaan J (2020) Modelling performance measurement barriers of humanitarian supply chain. Int J Prod Performance Manag. Vol. ahead-of-print No. ahead-of-print. http://doi.org/10.1108/IJPPM-01-2020-0031
37. Rugeviciute A, Mehrpouya A (2019) Blockchain, a panacea for development accountability? A study of the barriers and enablers for blockchain's adoption by development aid organizations. Front Blockchain 2(October):1–17. https://doi.org/10.3389/fbloc.2019.00015

38. Saberi S et al (2019) Blockchain technology and its relationships to sustainable supply chain management. Int J Prod Res 57(7):2117–2135. https://doi.org/10.1080/00207543.2018.1533261
39. Sander F, Semeijn J, Mahr D (2018) The acceptance of blockchain technology in meat traceability and transparency. Br Food J 120(9):2066–2079. https://doi.org/10.1108/BFJ-07-2017-0365
40. Schulz KA, Gstrein OJ, Zwitter AJ (2020) Exploring the governance and implementation of sustainable development initiatives through blockchain technology. Futures 122(July):102611. http://doi.org/10.1016/j.futures.2020.102611
41. Shahid A et al (2020) Blockchain-based agri-food supply chain: a complete solution. IEEE Access 8:69230–69243. https://doi.org/10.1109/ACCESS.2020.2986257
42. Shardeo V, Patil A, Madaan J (2020) Critical success factors for blockchain technology adoption in freight transportation using fuzzy ANP—modified TISM approach. Int J Inf Technol Decis Making. https://doi.org/10.1142/S0219622020500376
43. Shoaib M, Lim MK, Wang C (2020) An integrated framework to prioritize blockchain-based supply chain success factors. Ind Manag Data Syst 120(11):2103–2131. https://doi.org/10.1108/IMDS-04-2020-0194
44. Skidmore M et al (2017) The effect of intermediary market power on grain prices in India. In: 2017 agricultural and applied economics association annual meeting, Chicago, Illinois
45. Sushil (2012) Interpreting the interpretive structural model. Glob J Flex Syst Manag 13(2):87–106. http://doi.org/10.1007/S40171-012-0008-3
46. Sushil (2017) Modified ISM/TISM process with simultaneous transitivity checks for reducing direct pair comparisons. Glob J Flexible Syst Manag 18(4):331–351. http://doi.org/10.1007/s40171-017-0167-3
47. Trienekens JH et al (2012) Advanced engineering informatics transparency in complex dynamic food supply chains. Adv Eng Informatics 26(1):55–65. https://doi.org/10.1016/j.aei.2011.07.007
48. Tse D et al (2017) Blockchain application in food supply information security. In: 2017 IEEE international conference on industrial engineering and engineering management (IEEM), pp 1357–1361. http://doi.org/10.1109/IEEM.2017.8290114
49. Verdouw CN et al (2013) In: Kreowski H-J, Scholz-Reiter B, Thoben K-D (eds) Smart agri-food logistics: requirements for the future internet BT—dynamics in logistics. Springer, Berlin, pp 247–257
50. Wang Y, Han JH, Beynon-Davies P (2019) Understanding blockchain technology for future supply chains: a systematic literature review and research agenda. Supply Chain Manag 24(1):62–84. https://doi.org/10.1108/SCM-03-2018-0148
51. Wang Y et al (2019) Making sense of blockchain technology: how will it transform supply chains? Int J Prod Econ 211(April 2018): 221–236. http://doi.org/10.1016/j.ijpe.2019.02.002
52. Weber I et al (2016) In: La Rosa M, Loos P, Pastor O (eds) Untrusted business process monitoring and execution using blockchain BT—business process management. Springer International Publishing, Cham, pp 329–347
53. Yawar SA, Kauppi K (2018) Understanding the adoption of socially responsible supplier development practices using institutional theory: dairy supply chains in India. J Purchasing Supply Manag 24(2):164–176. https://doi.org/10.1016/j.pursup.2018.02.001
54. Zander K, Hamm U (2010) Consumer preferences for additional ethical attributes of organic food. Food Qual Prefer 21(5):495–503. https://doi.org/10.1016/j.foodqual.2010.01.006
55. Zhao G et al (2019) Blockchain technology in agri-food value chain management: a synthesis of applications, challenges and future research directions. Comput Ind 109:83–99. http://doi.org/10.1016/j.compind.2019.04.002
56. Zheng W et al (2019) NutBaaS: a blockchain-as-a-service platform, p 7

Chapter 8
Blockchain Technology for Enhancing Sustainability in Agricultural Supply Chains

Rohit Sharma, Anjali Shishodia, and Sachin S. Kamble

Abstract Nowadays, information and communication technology (ICT) plays a vital role in enhancing the sustainability performance in an organization. Technological disruptions such as the Industry 4.0 technologies have had significant positive impacts across various sectors, and agricultural sector is no exception. The current study highlights the impact of blockchain technology (BCT) as a sustainability enabler in the agricultural supply chains (ASCs). The study begins by identifying the BCT enablers from the literature that will lead to potential sustainability outcomes in the ASCs. These sustainability outcomes are then mapped with the United Nation's Sustainable Development Goals (UNSDGs). We therefore contend that BCT may hold the key in driving the ASCs toward sustainability. Finally, the possible implications from the study are suggested.

Keywords Blockchain · Sustainability · Agricultural supply chain

8.1 Introduction

Ever since blockchain technology (BCT) made its debut as a peer-to-peer transaction platform through the work of Nakamoto [29], it has revolutionized almost all the sectors wherever it has been deployed (see fintech, aerospace, manufacturing, tourism, supply chain management, and agriculture to name a few) [43]. Recent reports by PwC highlight that BCT through its potential use cases will add USD 1.76 trillion to the global gross domestic product (GDP) over the next decade [31]. Globally, BCT has garnered the attention of academia and practitioners through

R. Sharma (✉)
Jaipuria Institute of Management, Noida, India
e-mail: rohit.sharma.2017@nitie.ac.in

A. Shishodia
LMTSM, Thapar University, Patiala, India

S. S. Kamble
EDHEC Business School, Roubaix, France

R. S Mor et al. (eds.), *Operations and Supply Chain Management in the Food Industry*,
Lecture Notes in Management and Industrial Engineering,
https://doi.org/10.1007/978-981-16-5555-5_8

its fascinating use cases. BCT is a decentralized data management and transaction technology that eliminates the need of an intermediary. It is a platform that allows for the integration of other Industry 4.0 technologies such as digital twins (DT), cyber-physical systems (CPS), big data analytics (BDA), artificial intelligence (AI), augmented reality (AR), Internet of things (IoT), and Internet of services (IoS).

In the context of agribusiness, adoption of ICTs has tremendously helped increase the farm productivity and yield, and in mitigating the greenhouse emissions [4]. In addition, rapid globalization of the agri-food supply chains and the ever-growing world population have had an adverse impact on the sustainability of food supply chains [26]. Majority of the food production systems in the world are now competing for resources in order to ensure the food security for the masses. It is also estimated that majority of the food wastage and losses occur in the developing economies due to poor management of supply chain and technology integration issues [5]. The other challenges associated with the agricultural supply chains are seasonality, supply and demand spikes, long supply lead times, and perishability of agricultural produce [7]. Therefore, robustness and resilience are the major driving forces to manage agricultural supply chain risks. The research study carried out by Kamble et al. [18, 19] also highlighted critical issues in an agri-food supply chains such as lack of industrialization in the production processes, inadequacy of the top management, and information inaccuracy which render ASCs ineffective. The other issues threatening the agricultural supply chains are climate change risks, and development of sustainable agricultural supply chains is therefore one of the global challenges that needs to be overcome with the use of modern information and communication technology tools. Development of sustainable agricultural supply chains is therefore a global challenge that needs to be overcome. This involves identifying critical supply chain issues starting from the farm-level goals to global sustainability agenda. There has been increasing use of emerging technologies in the ASCs. Industry 4.0 technologies, IoT, BCT, and big data technologies are found to be the potential enablers for enhancing sustainability in the ASCs. These technologies are driving ASCs toward a digital supply chain environment which in turn would help in achieving sustainable supply chain objectives (social, environment, and economic).

In the present study, key BCT enablers and sustainable dimensions in ASCs have been identified from the literature which were then mapped to UNSDGs for examining their contribution in achieving the UNSDGs.

8.2 Literature Review

A literature review of the pertinent literature was carried out for identifying the BCT enablers in the ASCs. Only peer-reviewed articles published in Scopus were included in the review. Keywords for search included: blockchain, agricultural food chain, agricultural supply chain, and agri-food. These keywords were used independently and in combination with each other.

The literature review for the current study is divided into three subsections, viz. BCT and its enablers, ASCs and their sustainability dimensions, and UNSDGs.

8.2.1 BCT and BCT Enablers

Blockchains are digital ledgers that are tamper evident and tamper resistant and store information in a distributed, decentralized manner without the need of a central repository and a central authority [40]. Basically, BCT enables the stakeholders in a community to record the transactions on a shared ledger such that in case of normal operations, no transaction can be altered if once published [42]. BCT can be termed as a digital ledger, wherein all the transactions are stored in a list of blocks and new blocks are added as the information is appended on a continuous basis [44]. The BCT uses a decentralized database system, wherein the data is distributed simultaneously on different computers, and not stored on a single computer, that are known as nodes [15]. This unique characteristic of the BCT allows the stakeholders in a BCT network to access information on a real-time basis [25]. As BCT uses a cryptographic private key, data anonymity and data privacy of the network stakeholders/participants are thoroughly maintained, thereby making BCT safe and secure [24]. Some of the other prominent features of BCT are depicted in Table 8.1. BCT, therefore, holds immense potential in transforming traditional industries with its unique features such as decentralization, traceability, auditability, and transparency [18, 19].

8.2.2 Sustainability Dimensions in ASCs

Over the last few decades, the concept of sustainability in the agricultural supply chains has garnered a lot of interest from industry and the academia. As per the definition suggested by the World Commission on Environment and Development [38], sustainability is defined as "*the activities carried out in meeting the needs of the present without compromising the ability of future generations to meet their own needs.*" From the triple bottom line perspective [13], sustainability in the ASCs can be defined by the overlap of social, economic, and environmental dimensions.

The social sustainability dimensions in ASCs pertain to the development of short food supply chains promoting local products with an overall goal of inclusion of small landholders [17, 32]. The economic sustainability dimensions in the ASCs pertain to reducing the number of intermediaries in the ASC processes, reducing food frauds, and focusing on revenue sharing contracts between farmers and buyers for developing economically sustainable ASCs [17, 41]. The environmental sustainability dimensions of the ASCs pertain to activities that lead to food waste reduction, processes, and activities that increase food safety, security, and quality [33]. As there exists an overlap between the social, economic, and environmental dimensions, a

Table 8.1 Features of BCT

BCT enabler	Definition	Reference(s)
Real-time information flow	The presence of a decentralized and a distributed database system allows the stakeholders in a BCT network to access information on a real-time basis	Liu et al. [25], Mudliar et al. [28]
Decentralized database system	The data is distributed simultaneously on different computers known as nodes	Glaser et al. [15], Kouhizadeh and Sarkis [22]
Cybersecurity	BCT uses a cryptographic private key, and data privacy of the network participants is thoroughly maintained	Kshetri [24]
Traceability	Data can be traced back to each block via the time stamp	Kamble et al. [18, 19], Behnke and Janssen [6]
Transparency	The technology provides new frontiers in redesigning the reputation system and is helpful in preventing fraudulent activities	Chod et al. [12], Venkatesh et al. [37]
Immutability	The decentralized database technology allows the transactions to be stored in different nodes, thereby making it difficult to tamper with the data in the BCT network	Ateniese et al. [2], Zheng et al. [44]
Fast transaction settlement time	Decentralized and distributed database allows for faster transaction settlement times	Chiu and Koeppl [11], Peters and Panayi [30]
Reduced transaction time	BCT can unlock the sharing economy by making it cheaper to create and own an online platform. Transactions could be performed at lower costs	Min [27], Kamble et al. [18, 19]
Smart contracts	In a smart contract, the underlying asset is transferred into a program. This program then runs the code	Kosba et al. [21], Watanabe et al. [39]
Risk management	Deploying BCT ensures improved traceability of the underlying assets and adds to the provenance	Boussard et al. [8], Fu and Zhu [14]

balanced approach is required to be followed in order to implement BCT in the ASCs for achieving sustainable ASC operations.

The next subsection highlights the UNSDGs and how BCT supports in achieving them.

8.2.3 UNSDGs

The UNSDGs date back to 1992 Earth Summit, wherein more than one hundred seventy-eight countries adopted the Agenda 21 which was a detailed plan of action for building a global partnership for sustainable development aimed at improving human lives and protecting the environment. Following this, the member states adopted the 2030 agenda for sustainable development and provided a charter for inculcating peace and prosperity for the people and the planet. At the very crux of this charter are the seventeen Sustainable Development Goals (SDGs), which depict an urgent call for action by all member states both developed and developing to commit in a global partnership. These seventeen SDGs are aimed at ending poverty and other deprivations, developing strategies for improving health care and education, reducing inequalities, and enhancing economic growth [36]. In the case of ASCs, BCT is expected to bring in benefits supporting the UNSDGs. The targeted UNSDGs include UNSDG12 which is responsible production and consumption, UNSDG2 which is zero hunger, UNSDG3 which is good health and well-being, and UNSDG13 which is climate action.

The next section highlights the research framework adopted in this research.

8.3 Research Framework

Figure 8.1 depicts the research framework adopted for this research. The research consisted of four phases. In phase one of the research, the literature survey was executed for identifying the BCT enablers in the ASCs. In phase two of the research, the identified enablers were validated by the experts. Following the expert validation of the BCT enablers, these were then mapped with the sustainability outcomes, viz. social, economic, and environmental aspects of the ASCs in phase three. In the final phase, the sustainability outcomes of the ASCs were mapped with the UNSDGs based on the expert panel inputs.

The next section highlights the mapping of the sustainability outcomes of BCT with the SDGs. This mapping would help the academia and practitioners in designing sustainable ASCs using BCT.

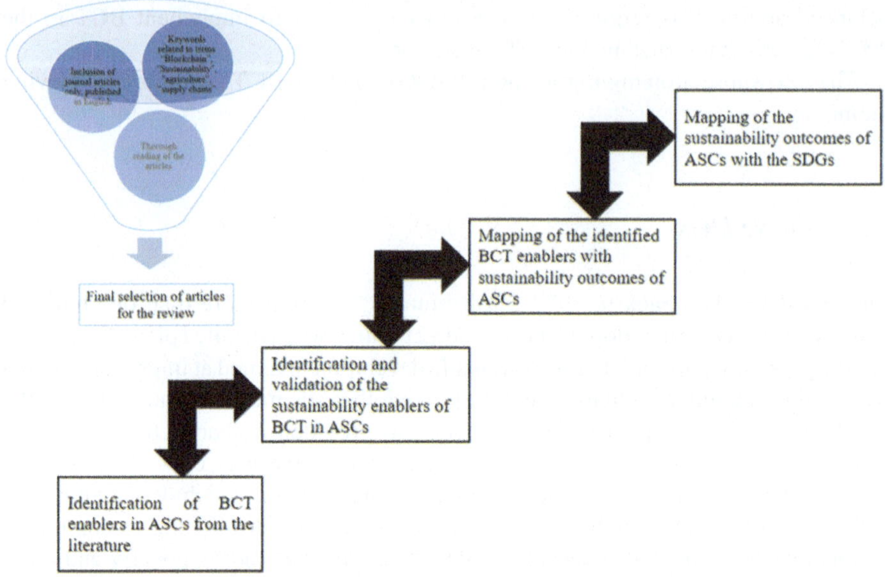

Fig. 8.1 Literature review process and research framework. *Source* Author

8.4 Mapping of Sustainability Outcomes with SDGs

For mapping the sustainability outcomes of ASCs enabled by BCT with the SDGs, opinion from an expert panel was sought. The expert panel comprised fifteen professionals (two blockchain experts; two environmental engineers; four senior-level practitioners from agribusiness organizations; four professors from academia having expertise in supply chain management, agricultural sciences, computer science, and information technology; two bank managers; and one system integrator with expertise in blockchain). Expert opinion reveals that BCT enablers can lead to potential sustainability outcomes in the ASCs. Figure 8.2 depicts the BCT sustainability framework, and it comprises three main components, viz. BCT enablers, ASC sustainability dimensions, and UNSDGs.

The first component of the BCT sustainability framework highlights the BCT enablers that have been found through extensive literature search. It is anticipated that the BCT enablers will have a profound impact in creating sustainable ASCs. The enablers impact all three sustainability dimensions, viz. social, economic, and environmental, thereby helping in creating sustainable ASCs globally. The second component of the BCT sustainability framework highlights the ASC sustainability dimensions. It can be highlighted that BCT will play a crucial role in efficient resource allocation for ensuring transparency and visibility in the ASCs [3]. BCT will further enable strategic information sharing and offers a promising future for achieving ASC sustainability [37]. The third component of the BCT sustainability framework highlights how the sustainability dimensions in the ASC can lead to the achievement

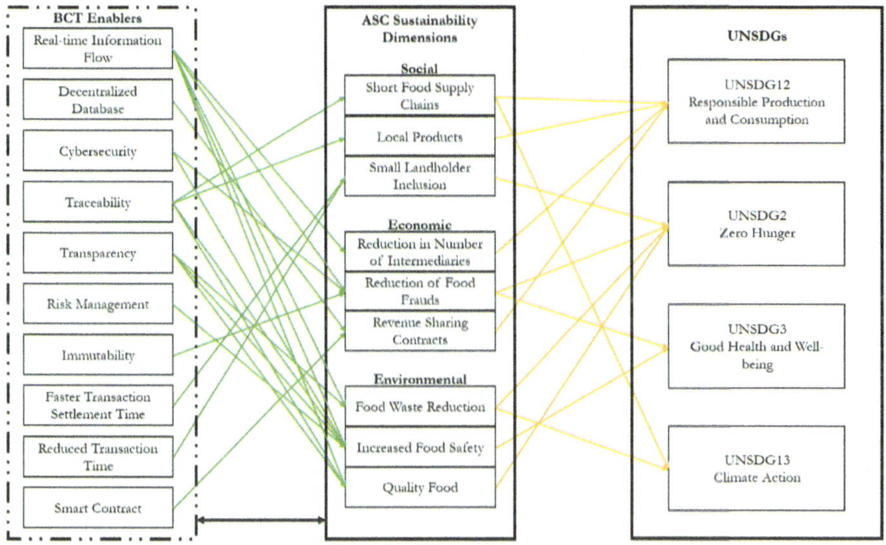

Fig. 8.2 BCT-sustainability framework. *Source* Author

of the UNSDGs. BCT will enable digitally enabled ASCs, thereby enhancing the transparency and visibility in the ASCs [35]. The BCT sustainability framework suggests that the BCT enablers and ASC sustainability dimensions have a feedback loop which focuses on developing the ICT capabilities for ensuring sustainable ASC operations through check and recheck mechanisms.

8.5 Discussion

BCT holds immense potential in transforming the traditional ASCs such as overcoming food fraud vulnerabilities and other counterfeiting issues [23]. It is therefore expected that incorporating BCT will enhance food safety and transparency through effective data capture, management, and control [9]. In line with the findings from the study carried out by Hughes et al. [16] relating BCT and the UNSDGs, the present study has mapped the BCT enablers with the sustainability outcomes of the ASCs for achieving the UNSDGs. Particularly, BCT helps in determining the ecological footprint by eliminating probable food frauds, promoting food safety, and reducing food waste across the ASCs. Overall, BCT will increase the environmental impact visibility in case of ASCs [1]. As reported by Chapron [10], BCT will enhance integrity of data through effective data management and control, thereby increasing the trust among ASC stakeholders. Overall, BCTs primary aim in ASC operations is to enhance the visibility and transparency through creation of digital ASCs, thereby promoting sustainable ASC operations which can help in achieving the targeted UNSDGs.

8.5.1 Implications

The agribusiness organizations should invest in developing sustainability orientation in the ASCs. The top management should ensure that adequate funds and resources are allocated for adoption of modern ICTs as these will enhance ASC sustainability. Moreover, agribusiness organizations should invest more in research and development for promoting innovation and sustainability. Usage of more sensors and devices along the ASCs will help in digitizing the ASCs and thereby enhance total quality management leading to sustainable ASCs [35]. The adoption of BCT in the ASCs will ensure effective and timely information sharing among the ASC stakeholders, thereby enhancing food traceability, food safety, and quality [34]. Top management should conduct training and development sessions (on modern ICTs and sustainability) for enhancing the knowledge and skillset of workers and other ASC stakeholders as it will change their outlook toward ASC sustainability [26]. With the adoption of BCTs in the ASCs, the practitioners will have a good food traceability system and they will be in complete control for managing food quality and safety [18, 19]. BCT adoption in the ASCs will enhance food provenance and will increase the ASC transparency, visibility, and traceability and thereby enhance the ASC sustainability.

As far as the policymakers are concerned, adoption and implementation of BCT will lead to effective regulatory and certification norms in the ASCs. This will further lead to reduction of fraudulent activities along the ASCs [20]. Further, BCT will help in facilitating the international trade by meeting all the sanitary and phytosanitary standards and requisite trade certifications/documentation. BCT will help in the financial inclusion of the farmers as the transaction settlement times will be reduced. Therefore, implementation of BCT in the ASCs will lead to social and economic sustainability of farmers and small landholders. Further, BCTs can lead to the elimination of middlemen and ensure direct credit facilitation for the farmers. Features of BCT such as auditability and immutability will help in maintaining an efficient audit trail. BCT will effectively help in mitigating the risks in the ASCs (pertaining to food safety, quality, transaction settlement, credit facilitation, and insurance services) and help in enhancing the resilience capabilities in the ASCs leading to creation of sustainable ASCs [18, 19].

8.5.2 Limitations and Future Research Directions

The present study identified ten BCT enablers for enhancing the sustainability in the ASCs. These enablers were identified from the pertinent literature and were validated by experts from various fields such as agribusiness, supply chain management, computers, and information technology. BCT in particular has certain limitations that need to be overcome so as to reap the benefits of this technology. A few notable shortcomings of BCT are its high implementation cost, scalability issues, the need to skill the current workers with BCT, and interoperability issues. As the study pertained to a developing economy context, enablers might have been different in developed economies. Therefore, it is recommended to carry out similar studies in the context of developed economies as their ASCs are comparatively advanced and the results may then be compared for further refinement of the interdependencies. BCTs are still being tested on experimental basis; therefore, case studies can be carried out for highlighting the significant work the agribusiness organizations have been doing to enhance sustainability in the ASCs using BCT. Future studies can be carried out based on finding the cause and effect relationships between the BCT enablers using structural equation modeling and other modeling techniques.

References

1. Ahmed S, ten Broek N (2017) Blockchain could boost food security. Nature 550(7674):43–43
2. Ateniese G, Magri B, Venturi D, Andrade E (2017) Redactable blockchain-or-rewriting history in bitcoin and friends. In: 2017 IEEE European symposium on security and privacy (EuroS&P). IEEE, pp 111–126
3. Bai C, Sarkis J (2020) A supply chain transparency and sustainability technology appraisal model for blockchain technology. Int J Prod Res 58(7):2142–2162
4. Balafoutis A, Beck B, Fountas S, Vangeyte J, Wal TVD, Soto I, Gómez-Barbero M, Barnes A, Eory V (2017) Precision agriculture technologies positively contributing to GHG emissions mitigation, farm productivity and economics. Sustainability 9(8):1339
5. Balaji M, Arshinder K (2016) Modeling the causes of food wastage in Indian perishable food supply chain. Resour Conserv Recyc 114:153–167
6. Behnke K, Janssen MFWHA (2020) Boundary conditions for traceability in food supply chains using blockchain technology. Int J Inf Manage 52:101969
7. Behzadi G, O'Sullivan MJ, Olsen TL, Zhang A (2018) Agribusiness supply chain risk management: a review of quantitative decision models. Omega 79:21–42
8. Boussard M, Papillon S, Peloso P, Signorini M, Waisbard E (2019) STewARD: SDN and blockchain-based trust evaluation for automated risk management on IoT devices. In: IEEE INFOCOM 2019-IEEE conference on computer communications workshops (INFOCOM WKSHPS). IEEE, pp 841–846
9. Bumblauskas D, Mann A, Dugan B, Rittmer J (2020) A blockchain use case in food distribution: do you know where your food has been? Int J Inf Manag 52:102008
10. Chapron G (2017) The environment needs cryptogovernance. Nature 545(7655):403–405
11. Chiu J, Koeppl TV (2019) Blockchain-based settlement for asset trading. Rev Fin Stud 32(5):1716–1753
12. Chod J, Trichakis N, Tsoukalas G, Aspegren H, Weber M (2019) On the financing benefits of supply chain transparency and blockchain adoption. Manag Sci 66(10):4378–4396
13. Elkington J (1998) Partnerships from cannibals with forks: the triple bottom line of 21st-century business. Environ Qual Manage 8(1):37–51
14. Fu Y, Zhu J (2019) Big production enterprise supply chain endogenous risk management based on blockchain. IEEE Access 7:15310–15319
15. Glaser F, Hawlitschek F, Notheisen B (2019) Blockchain as a platform. In: Business transformation through blockchain. Palgrave Macmillan, Cham, pp 121–143

16. Hughes L, Dwivedi YK, Misra SK, Rana NP, Raghavan V, Akella V (2019) Blockchain research, practice and policy: applications, benefits, limitations, emerging research themes and research agenda. Int J Inf Manage 49:114–129
17. Jelsma I, Schoneveld GC, Zoomers A, van Westen AC (2017) Unpacking Indonesia's independent oil palm smallholders: an actor-disaggregated approach to identifying environmental and social performance challenges. Land Use Policy 69:281–297
18. Kamble SS, Gunasekaran A, Gawankar SA (2020) Achieving sustainable performance in a data-driven agriculture supply chain: a review for research and applications. Int J Prod Econ 219:179–194
19. Kamble SS, Gunasekaran A, Sharma R (2020) Modeling the blockchain enabled traceability in agriculture supply chain. Int J Inf Manag 52:101967
20. Kamilaris A, Fonts A, Prenafeta-Boldú FX (2019) The rise of blockchain technology in agriculture and food supply chains. Trends Food Sci Technol 91:640–652
21. Kosba A, Miller A, Shi E, Wen Z, Papamanthou C (2016) Hawk: the blockchain model of cryptography and privacy-preserving smart contracts. In: 2016 IEEE symposium on security and privacy (SP). IEEE, pp 839–858
22. Kouhizadeh M, Sarkis J (2018) Blockchain practices, potentials, and perspectives in greening supply chains. Sustainability 10(10):3652
23. Kshetri N (2018) 1 Blockchain's roles in meeting key supply chain management objectives. Int J Inf Manage 39:80–89
24. Kshetri N (2017) Can blockchain strengthen the internet of things?. IT Prof 19(4):68–72
25. Liu H, Zhang Y, Yang T (2018) Blockchain-enabled security in electric vehicles cloud and edge computing. IEEE Netw 32(3):78–83
26. Mangla SK, Luthra S, Rich N, Kumar D, Rana NP, Dwivedi YK (2018) Enablers to implement sustainable initiatives in agri-food supply chains. Int J Prod Econ 203:379–393
27. Min H (2019) Blockchain technology for enhancing supply chain resilience. Bus Horiz 62(1):35–45
28. Mudliar K, Parekh H, Bhavathankar P (2018) A comprehensive integration of national identity with blockchain technology. In: 2018 international conference on communication information and computing technology (ICCICT). IEEE, pp 1–6
29. Nakamoto S (2009) Bitcoin: a peer-to-peer electronic cash system. Bitcoin. org. Disponible en https://bitcoin.org/en/bitcoin-paper
30. Peters GW, Panayi E (2016) Understanding modern banking ledgers through blockchain technologies: future of transaction processing and smart contracts on the internet of money. In: Banking beyond banks and money. Springer, Cham, pp 239–278
31. PwC (2020) https://www.pwc.com/gx/en/news-room/press-releases/2020/blockchain-boost-global-economy-track-trace-trust.html. Last accessed 2020/11/10
32. Schmitt E, Galli F, Menozzi D, Maye D, Touzard JM, Marescotti A, Six J, Brunori G (2017) Comparing the sustainability of local and global food products in Europe. J Clean Prod 165:346–359
33. Sgarbossa F, Russo I (2017) A proactive model in sustainable food supply chain: insight from a case study. Int J Prod Econ 183:596–606
34. Tian F (2016) An agri-food supply chain traceability system for China based on RFID & blockchain technology. In: 2016 13th international conference on service systems and service management (ICSSSM). IEEE, pp 1–6
35. Tsolakis N, Niedenzu D, Simonetto M, Dora M, Kumar M (2021) Supply network design to address United Nations sustainable development goals: a case study of blockchain implementation in Thai fish industry. J Bus Res 131:495–519
36. UNDESA (2020) United Nation's Department of Economic and Social Affairs sustainable development. https://sdgs.un.org/goals
37. Venkatesh VG, Kang K, Wang B, Zhong RY, Zhang A (2020) System architecture for blockchain based transparency of supply chain social sustainability. Robot Comput Integr Manuf 63:101896

38. WCED (1987) Our common future, vol 17. World Commission on Environment and Development, pp 1–91
39. Watanabe H, Fujimura S, Nakadaira A, Miyazaki Y, Akutsu A, Kishigami JJ (2015) Blockchain contract: a complete consensus using blockchain. In: 2015 IEEE 4th global conference on consumer electronics (GCCE). IEEE, pp 577–578
40. Yaga D, Mell P, Roby N, Scarfone K (2019) Blockchain technology overview. arXiv preprint https://arxiv.org/abs/1906.11078. https://nvlpubs.nist.gov/nistpubs/ir/2018/NIST.IR.8202.pdf
41. Yan B, Shi S, Ye B, Zhou X, Shi P (2015) Sustainable development of the fresh agricultural products supply chain through the application of RFID technology. Inf Technol Manage 16(1):67–78
42. Yli-Huumo J, Ko D, Choi S, Park S, Smolander K (2016) Where is current research on blockchain technology?—a systematic review. PloS one 11(10):e0163477
43. Zhang P, Schmidt DC, White J, Lenz G (2018) Blockchain technology use cases in healthcare. Adv Comput 111:1–41
44. Zheng Z, Xie S, Dai H, Chen X, Wang H (2017) An overview of blockchain technology: architecture, consensus, and future trends. In: 2017 IEEE international congress on big data (BigData congress). IEEE, pp 557–564

Chapter 9
Cloud Computing: Technological Innovations in the Food Industry

Kumar Rahul, Rohitash Kumar Banyal, Vijay Kumar, and Darshana Hooda

Abstract Cloud computing is an emerging area of computer sciences, and much research is in progress in different aspects since the beginning. Cloud computing creates security, cost, and governance concerns in other prospects in use and a network system of food industries. Food industries are the fastest growing industries in India and abroad. Simultaneously, cloud computing plays an essential role in data and application migration to the cloud for better and effective services. Enterprises get multiple benefits in terms of hardware and software through cloud computing technologies. This book chapter describes various aspects of cloud computing implementation challenges and issues among distributed systems that are well connected geographically. This chapter also provides multiple levels of challenges during the implementation of cloud services in the food industries at the server-side and client-side. This chapter describes a systematic literature review of benefits, technologies, challenges, and issues encountered during cloud computing services offered to food industries.

Keywords Cloud computing · Scalability · Security · Food industries · Services

BDA	Big data analysis
CDC	Cloud data center
DAAS	Desktop as a service
ERP	Enterprise resource planning
GUI	Graphical user interface

(continued)

K. Rahul (✉) · V. Kumar
Department of Basic and Applied Science, National Institute of Food Technology Entrepreneurship and Management, Kundli (Sonepat) 131028, India

R. K. Banyal
Department of Computer Sciene and Technology, Rajasthan Technical University, Kota 321040, India

D. Hooda
University Computer Center, Deenbandhu Chhotu Ram University of Science and Technology, Murthal 131039, India

R. S Mor et al. (eds.), *Operations and Supply Chain Management in the Food Industry*, Lecture Notes in Management and Industrial Engineering,
https://doi.org/10.1007/978-981-16-5555-5_10

(continued)

IAAS	Infrastructure as a service
IoT	Internet of things
NAAS	Network as a service
PAAS	Platform as a service
SAAS	Software as a service
SBMC	Subscription basis model charges
TMS	Transaction management system

9.1 Introduction

Cloud computing is a technology, mechanism, or approach that provides different services regarding hardware and software to the various clients geographically located with a limited computing power capacity at their end. Cloud computing offer services like manipulation, configuration, and accessing application online. Cloud computing provides services like data storage facilities, infrastructure setup facilities, and application access online. Various cloud computing applications exist, including higher education, health, banking, retail, manufacturing, transport, pharmacy, and service sector. These sectors are well equipped with automation and systematic execution of their operations with the IT facilities. Cloud computing consists of various attributes such as multi-tenancy, elasticity, network access, pay as you go, and on-demand resources [1]. Industrial data are essential data for their daily operations and maintenance; their application and usability are different regarding profitability. Food industries' IT support cannot be changed again and again in a shorter timeline. Thus, a facility called cloud computing helps to "pay for the services they have used." Cloud computing provides services at the level of hardware through Software as a Service (SAAS), Platform as a Service (PAAS), and Infrastructure as a Service (IAAS) accordingly. Nowadays, approximately 90% of global industries use cloud computing as a service and make business more profitable. A significant feature of cloud computing is multiple tenant-oriented services, meaning different service users can run other platforms or software concurrently on the same infrastructure.

Maximum utilization of cloud resources is the primary concern of the client system or industries always provide a better outcome, and cloud resource is an essential requirement for successful implementation at the operational and execution level [2]. Observations have been made for cloud adoption in industries, which leads to the study of different sectors. The level of service access has been different from GUI-based software to the infrastructure levels. If we talk about the banking sector located at other places and remote locations, service providing facilities has always been challenging in less time and money. We all know these sectors are well connected to e-com industries, where transaction failure affects industry outcomes [3]. However, cloud in the manufacturing industry, only 7% believe that cloud services are applicable effectively [3]. Recently, mobile cloud computing has emerged in this area and shown how mobile cloud computing is used. Mobile cloud computing offers benefits such as scalability, reliability, battery life, etc. but still has lots of challenges such as privacy, trust, energy efficiency, security, bandwidth, data management, synchronization [4]. Cloud computing characteristics include on-demand self-service, broad network access, rapid elasticity, measured service, resource pooling, etc. [5].

Food and beverage industries can use cloud computing technologies such as PAAS, NAAS, IAAS, SAAS, and so on a different scale. Food industries require delivery of the product on time through supply chain management technologies, enabling cloud services such as "SAAS" effectively. Supply chain management software can be helpful through SAAS. Cloud supply chain (SAAS, IAAS, PAAS) integrated with suppliers, distributors, manufacturers, wholesalers, retailers [5]. The "on-time food delivery system" challenge can be solved with the cloud computing technological implementation concept in food industries. By doing so, food industries would be in a position to have higher benefits. Cloud computing services enable food industries to accelerate the operations and functionalities along with supply chain objectivities.

The different tasks such as traceability, cost-effectiveness, streaming complexity, and data sharing among various end-users of food industries can be achieved through cloud computing services [5]. 74% of food and beverage industries have adopted cloud computing as per Harvard business review claimed in Operators [6]. The food and beverage sector would benefit from adopting cloud computing services in different ways, including data security, reduction of time and cost, decision support system improves, control staff and inventory cost, consistency through centralization, and compliance with regulations [6]. This book chapter is divided into different sections. Section 9.2 describes cloud computing in terms of industrial applications, which defines the basic understanding and requirement of the cloud for industries and how much (%) different industries are using cloud technologies in today's scenario. Cloud computing is suitable for the food industries for various reasons including greater speed and agilities, updating system within food industries, expanding locations with lower cost, executing new ideas, interaction with customers through the latest communications mode, revenue potential increase, decision-making system with big data technologies and so on [6].

In Indian food industries, cloud computing technologies have been implemented and executed by a few legendary industries such as Nestle, Haldiram's, etc. It is estimated that by 2022, more computation services provided by PAAS and IAAS [7]. Cloud computing services help in storage, processing, and networking through artificial intelligence and machine learning techniques [7]. The next industrial growth would be affected by cloud computing technologies. Around 60% of the enterprises (including food industries) would be in a position to use cloud services at the end of 2020 [7]. Indian cloud computing services report 2020–2027 indicates the implementation of cloud services requirements in small- and medium-scale industries [8]. Cloud services providers are Amazon web services, Azure, Google cloud, Alibaba cloud, IBM cloud, Tencent cloud, Oracle, etc. Amazon leads to $100 Billion cloud market as per records given in synergy research group [9].

The worldwide public cloud services market has grown up to $257.9 billion as per Gartner, Inc. [10]. DAAS is suitable for remotely located employees of an organization. It has significant growth from 95.4% to $1.2 billion and is expected worldwide public cloud services for different services IAAS, SAAS, and PAAS [10]. Cloud migration services is expected to be $448.34 Billion (USD) by 2025. It has been increasing from smaller to large-scale organization for scalability, effectiveness and disaster recovery, etc. In the recent pandemic of COVID-19, most industries have moved to cloud services with a witness of 775% cloud services claimed by Microsoft Inc. [11]. The cloud migration market growth rate by region (2020–2025) shows North America is the leading one, where 95% of organizations have moved toward cloud services, a survey claimed by IBM [11]. Cloud computing adoption maturity of the UK has given in Opara-Martins et al. [12]. In the survey paper, it is mentioned, 46.2% of industries preferred to migrate and use public cloud, whereas 36.8% industries used hybrid cloud, followed by 26.4% industries in private cloud and merely 10.4% industries only for community cloud [12].

Section 9.2 explains cloud computing: an overview in terms of food industries applications. Section 9.3 describes the benefits of cloud computing for food industry solutions. Section 9.4 presents cloud computing implementation and challenges for food industries. Section 9.5 focused on the outcome for food industries after cloud concept (services) implementation. Section 9.6 elaborates on various issues and the future of cloud computing in food industries for better and effective utilization in industries.

9.2 Cloud Computing: An Overview in Terms of Food Industries Applications

There is an increasing demand for cloud computing usability in many industries [13]. Cloud computing provides services through various network mode. Cloud computing structure presented in scalable infrastructure supports different applications, accessing files, or data storage globally. With these technologies, multiple costs such as computational, data storage, website hostage, application hosting, and service delivery reduce. Cloud computing is based on providing service of reusability and effective utilization of IT services [14]. Different industries, distinct in nature and production, use cloud computing for business continuity plan (BCP) and disaster recovery purposes. Nowadays, industrial services are cloud-based/IT-based, which is not always possible to be available effectively in industries. They (industries) benefit from SAAS, and for all such facilities, industries pay to the cloud service providers. A well-deployed cloud computing service provides many advantages to sectors regarding easy access to resources and applications, including data [15]. It offered cost-effectiveness and built up to standard services among various cloud service providers. Cloud computing adoption in industries reduces capital and operational expenditure [15].

Food industries tend to understand that scalability and availability are possible to achieve [15]. The authors showed scheduling in the cloud and scheduler organization for a distributed system where the scheduler input and output data, scheduler runs, frequency, application model, and objectives [7]. Here, the authors show the client (as industries) and provider (cloud service) perspective in cloud computing [16]. The author shows interrelated issues, emerging trends, and challenges near IoT, big data, and cloud computing in all respect [17]. The author said IoT and big data more influence cloud computing technologies future growth [17]. The author shows the adoption of cloud computing in e-government and discussed the cloud computing role advanced from the traditional government system concerning the feature of availability, accessibility, performance, data scalability, and disaster recovery industries [18]. Cloud computing characteristics include on-demand self-service, multi-tenancy and resource pooling, rapid elasticity, scalability, resource optimization, etc. [19].

9.2.1 *Cloud Computing Technologies of Food Industries*

Cloud computing reduces cost, reduces management responsibilities, increases business agility, and meets enterprises flexible environment [19]. It is suitable for the whole logistics platform, as mentioned in Daniluk and Holtkamp [20]. It supports logistics processes and value-added services to the logistics sector to boost up productivity. The cloud computing market value has increased from $25.5 billion to $159.3 billion in 2020 [20]. The authors have designed a logistics process designer (LDP), a central tool for the logistics mall [20]. Cloud computing is also called utility computing for industries. Cloud technologies are used to provide services of different categories of services to various industries for IT solutions, including banking, retail, manufacturing, etc. Cloud technologies include:

i. *Commodities of Hardware*
ii. *Virtualization of Computational Processing, Storage and Communication*
iii. *Distributed System Services.*

Industries have started using cloud computing, easily understood from the diagram (Fig. 9.1). Various sectors such as insurance, oil, gas and electric, school and education, food retail and healthcare, financial sector, manufacturing, telecommunication, business management services, specialized services, freight services, energy management, membership organization, media, and military use cloud computing services. Effective utilization of cloud computing in various sectors can be understood as given below.

Different industries (given below in Fig. 9.2) use cloud computing technologies for other purposes. The cloud delivers SAAS, PAAS, and IAAS, and so on to many clients as and when required. It is an on-demand service system. It includes the latest technologies such as virtualization, live migration (LM), offline access, online storage, third-party integration, online collaboration, etc. It offers applications of variance in nature with high computing parameters. These services shifted to the

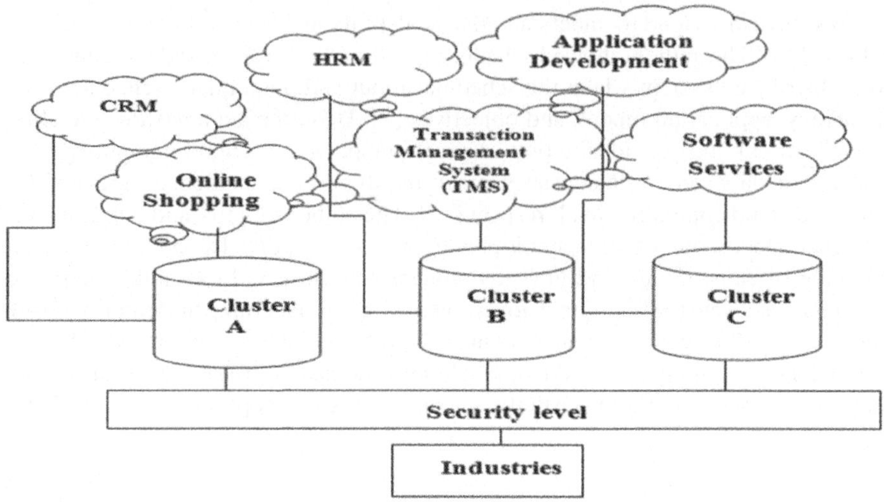

Fig. 9.1 Cloud computing applications. *Source* Author

outsourced concept under industries for recommendations, including the cost of hardware/software setup and IT administration. The industrial security level consists of various clusters of different natures where groups provide application-based software and other essential applications for industries at different levels.

Figure 9.1 shows Cluster A offers services for online shopping related web application, which also includes Customer Relationship Management (CRM) module for maintaining the relationship with the customer to promote other product promotions through the web portal. Cluster B maintains the dataset of various TMS users, which keeps the record of, "How many transactions occurred during a particular date and time?" Cluster C keeps records of "Requirement of applications software for industries based on demand." Cloud computing service provider industries such as Amazon, IBM Cloud, Oracle Cloud, SAP, Red Hat, Verizon Cloud, etc., provide these services to various industries.

9.2.2 Cloud Computing in Different Industries

Various industries are using big data and cloud computing technologies for the last few years. Day by day, their industrial growth is increasing, and at the same time, it becomes essential for them to maintain quality services and produce the right product on time. Figure 9.2 shows cloud usages in various industries. It represents how much % of different industries are using cloud services. It offers financial sector 12%, manufacturing 10%, telecommunication 9%, business management 10%, specialized service 5%, energy, membership organization 6%, media and military 6%, chemical and pharmacy 3%, commercial physical research 1%, government 7%, insurance

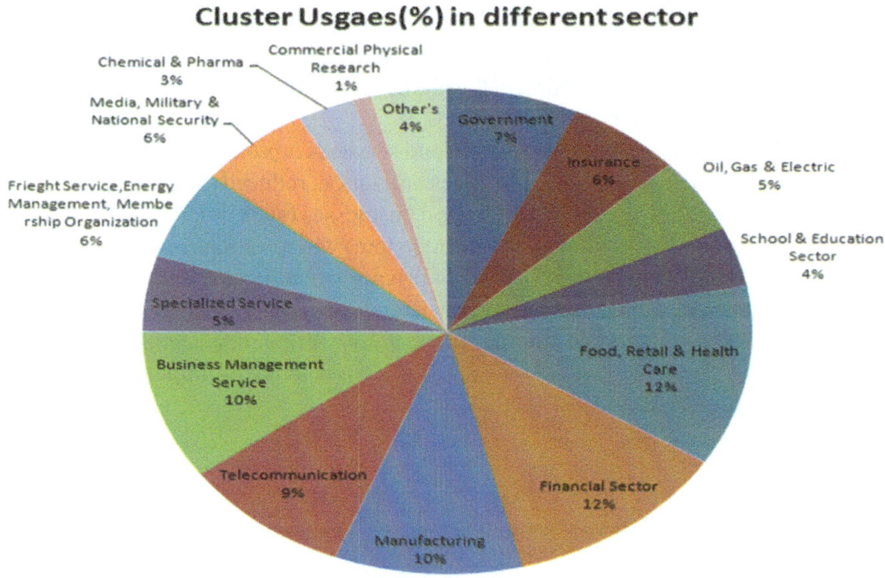

Fig. 9.2 Cloud usages of various industries [21]

6%, oil gas and electric 5%, school and education 4%, food retail and healthcare 12%, and another sector 4%. In food retail and healthcare, the food habit is affected by a celebrity chef in different ways at a different level with the support of media technologies [21].

9.3 Benefits of Cloud Computing for Food Industries Solutions

There are many benefits to using cloud computing in industries. Industries have been using to solve operational problems and to reduce cost optimization. Cloud computing benefits can be understood in terms of easily upgraded software, data storage, reduce IT cost, business continuity, productivity anywhere, efficiency, scalability, flexible cost, always-on availability, improved mobility, and improved collaboration. Another advantage of cloud computing is the minimum cost of globalization of the workforce. Industries (other than IT) do not require to spend more on IT infrastructure, including software upgradation and software and network licensing. With the help of a login/password or accessibility feature, industrial data can be access across the globe.

Other benefits of these technologies are providing project monitoring and controlling systems that work effectively, rapid development, high productivity, less deployment time, easy to manage, high availability, and shared resources. Cloud computing services and IoT are available to support cultivation, agri-logistics, food processing, distributor, food retail, food services, etc. [22]. In cold chain logistics, cloud computing and BDA work to estimate the predictions and real-time traffic congestion and transportation through unified device [23].

Through data-center consolidation of cloud services, it is easier for food industries to achieve lower cost for computer resources utilization, reduced time to market, etc. It supports multi-tenancy for the different food businesses (such as ERP, food information system, traceability system, production system) where various operations will be useful and maintained separately in form of privacy isolation, performance isolation, failure isolation [24]. Cloud computing services also support statistical multiplexing, horizontal scalability, variable pricing, price setting, efficiency, incentive compatibility, etc. [24]. However, these three are significant benefits of implementing cloud computing in food industries as follows.

9.3.1 Reduce IT cost

However, cloud computing provides IT services based on demand anywhere. Hence, no need to maintain and expense more on IT establishment and infrastructure much in industries other than IT, only up to a certain level, enterprises need to purchase an established IT infrastructure, flexibility offers in IT cases, where software or platform is used directly from the cloud. There is no need to pay wages to external experts. Since industries have limited IT resources, current IT infrastructure forces them to focus more on data storage, security, retrieval issues, etc. Cloud computing services provider industries offer more IT-enabled services.

9.3.2 Business Continuity

Business continuity is an essential feature of cloud computing in industrial aspects. It is basically a "plan of execution" in case of business failure. The BCP objective (business continuity plan) is "Business should not stop, in any case". So, the data, process, engineering methodology, production steps, quality parameter, cost evaluation, etc., everything stored on the cloud for fast transfer in case of emergency. Business-related data stored on a cloud storage system, so data security is an essential aspect for cloud computing service provider companies. Cloud service provider company provides you data as and when required to loss of productivity and back-end support.

9.3.3 *Software Integration*

Cloud computing is nicely used through different ERP systems. ERP software is implemented in almost all types of industries.

The majority of the ERP is integrated with other marketing, production, supply chain, advertisement, human resource, customer relationship management, etc. SAAS services are readily available with cloud technology, and with the help of SAAS, the software is integrated and upgraded on time. More enterprises are giving first preference to cloud services in 2018 as per the report of RightScale [25].

9.4 Cloud Computing Implementations Challenges and Opportunities for Food Industries

Cloud computing challenges are categorized in different ways, which is given in this section. However, it includes a lack of resources and expertise, security, managing cloud, compliance, government, etc. Cloud computing challenges can be understood from the diagram mentioned below [25].

Various implementation challenges need to be resolved before implementation and access resources of cloud services.

i. *Identifying the suitable cloud setup*: Suitable Cloud setup is an essential step for any business or industry to meet their requirements. Cloud setup ordinarily available in different forms such as private cloud, community cloud, and hybrid cloud. All these clouds are other in size, but functionalities are the same for execution and services. Every industry has its own set up for IT needs. All efforts made to meet their specific requirements to run a business such as beverage

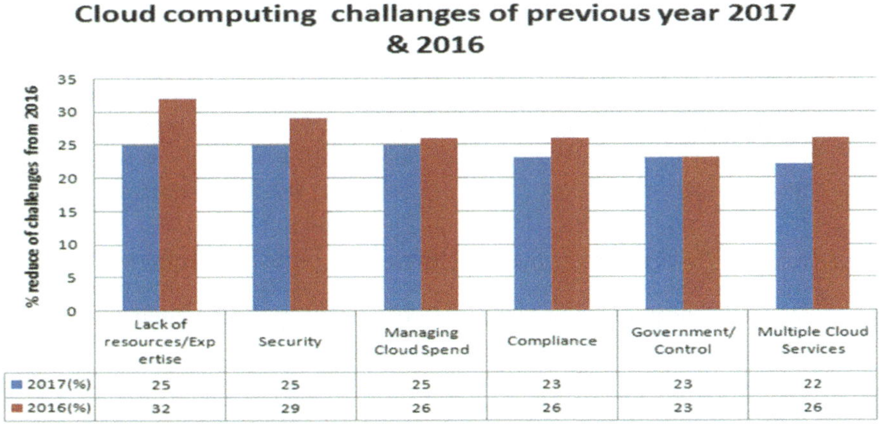

Fig. 9.3 Cloud computing challenges in the previous year of 2017 and 2016 [25]

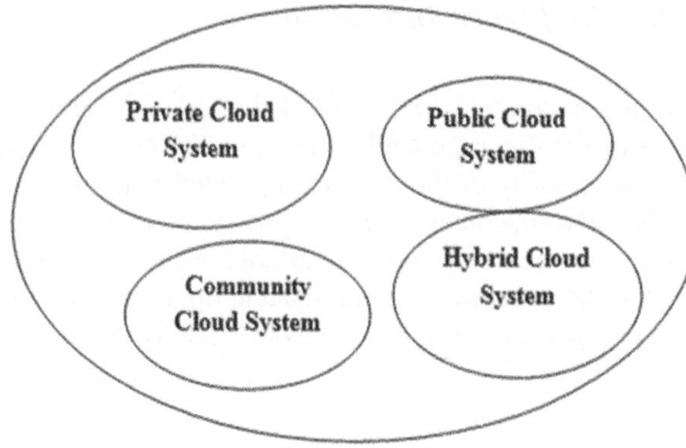

Fig. 9.4 Cloud computing and its type. *Source* Author

industries will focus more on beverage product making process, leveling, quality evaluation, packaging, and marketing rather than IT services. However, there would be a need to utilize the right cloud services in the automation system to achieve the success rate without establishing more IT facilities in industries [26].

ii. *Unauthorized service provider*: Unauthorized service provider is the second important security and services of cloud computing. Industries have ICT facilities to cope up with the requirement and economic growth of enterprises. Figure 9.1 shows different cloud computing concept and their types of implementations, so industries can adopt anyone to provide IT facilities in various forms with secured and authorized services. The majority of the sectors (except IT) does not put much effort into ICT facilities instead, and they focus on core areas to gain maximum profit. Thus, it becomes essential to understand the authorized service provider from the cloud side.
iii. *Data security*: Data security is an important parameter, and it is one of the biggest challenges for industries in different forms to provide quality services. Industrial data are sharable among various units in an organization where data access permits any other industries to use at a certain level, which must be protected and secure during transmission. There have been many malicious attacks in the past, including virus attacks, threats, hacking at micro and significant access levels in industries that force us to think about data security in sectors. That is why entrepreneurs feel about security issues while implementing cloud services and accessing cloud services in industries [26]. Industrial policies, processing strategy, product making process, etc., are essential aspects of industrial data, so organizations don't share without permission and provide access. So, cloud computing service provider companies' responsibilities are to ensure that their "data will be secure" and not shared among third parties.
iv. *Consumptions basis charges*: This is also called consumption-based model charges (CBMCs), which charge as per consumptions and services. Cloud-provided services as on-demand, so sometimes it quite happens that price goes up and down as per availability of resources. It is not possible to stick always

to a particular amount. Because of this reason, it is difficult to implement cloud services in industries. Since this is a pricing model, there is a subscription model available at the counter that provides web services up to a certain level, whether daily, monthly, yearly, etc. Nowadays, such a subscription is easily available in different sectors, such as the telecom sector (BSNL, MTNL, RELIANCE, etc.). This approach is called subscription basis model charges (SBMCs), where customers need to pay as per time frame and subscription limit.

v. *Migration:* Industry's high authority cultural difference is one of the issues to shift from one platform to other platforms such as access services on payment based on cloud computing technologies. Industries high authorities generally do not share essential data details to third party or through technologies intermediate devices.

   Higher authorities are having misconceptions about cloud computing (except IT industries) because it is hard for them to accept technological advancement in most industries [26].

vi. *Reliability on Cloud Computing Services*: Not all industries believe in cloud computing facilities and services easily available on-demand where no-cost implementation exists for hardware and software respect. In fact, with the various feature, work becomes easy and reduce overhead cost. Industrial data transformation from living to a cloud system is difficult because of the lack of reliability on technologies [27].

   With the feature advancement of cloud computing services, the CDC increases their size and volume, where resource failure occurs. Thus, industries have less confidence in the reliability of services provided by the cloud [27].

vii. *Transforming data to virtual setup*: Transforming data and business from one part to another virtual form is difficult for any industry, where some portion of enterprises is carried out at one place, and another amount of work will execute at virtual design.

9.5 The Outcome for Food Industries After Cloud Services Implementation

Cloud computing-based applications in industries do not pay for different hardware and software facilities to avail. However, SAAS, PAAS, and IAAS of cloud technology provide all services related to software, platform, and infrastructure. Food and beverage industries IT cost reduced after implementation of cloud services. Food industries and IT infrastructure reduced. Services are being upgraded automatically when it is being used through cloud activities. It is also upgraded periodically when interacted with cloud services or technology hired by the cloud. Cloud adoption leads to less deployment time and effective utilization of resources at an industrial

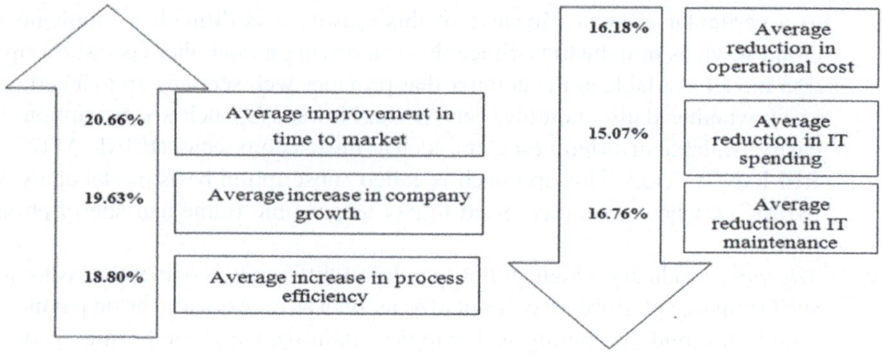

Fig. 9.5 Impact of cloud computing in food industries [28]

site. Instead, cloud adoption supports adopting more projects and work related to IT services and free from purchase in industries (Fig. 9.5).

The above figure shows the cloud impact on business [28]. Industries that are equipped with cloud services experienced 20.66% average improvement in time to market, 18.80% recorded as moderate growth in process efficiency within an organization, and a 15.07% reduction in IT spending of industries. All together, these sources led to a 19.63% increase in industrial growth. In the case of public cloud services adoption among sectors, the following observations were found as per the report given [25].

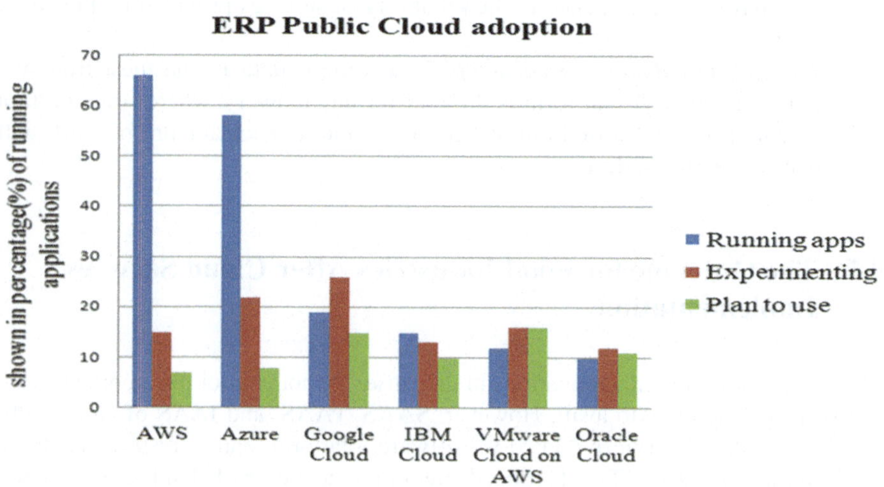

Fig. 9.6 ERP public cloud adoption [25]

9.6 Issues and Future of Cloud Computing in Food Industries

Some challenges faced by Indian food industries such as poor knowledge of weather forecasting, pest and diseases, poor ICT infrastructure, insufficient power supply in rural areas where small- and medium-scale industries installed and function [29]. So, these challenges can be solved through cloud computing services with a minimal rental cost. Industries would be able to get on-demand, less expensive, measured services, etc. Cloud-based services provides databases for crop related information, market information, agricultural tools and production information [29]. Some other challenges include IoT under development, capital cost, skilled workforce, farm monitoring devices, advanced farm equipment, alternate farming techniques, within food industries.

Although there are various issues and challenges exist during the cloud computing adoption in industries such as security (lack of control over the location of data, lack of filtered data, increased vulnerabilities from shared infrastructure, privileged use abuse at the cloud provider), reliability, stability, access, interorganizational connectivity, and application software availability [30]. Apart from these challenges, industries face several other challenges during cloud computing technology implementation in enterprises.

These challenges are integration with existing systems, modifying workflow, changes to IT policy, learning curve, higher cost than expectation, lower performance than expected, lower availability than expected, and so on [31]. The future of cloud computing affected food industries in terms of enhancing storage services and the Internet, ERP development preferred in modular form, IoT emerged with cloud services, the predictive mechanism through big data, update cloud services, etc. Food industries also face several other issues, including data security risks, over dependencies on a single cloud, legal and regulatory compliances, data access, lack of integration between modules and network, data protection and security, transition cost are high, loss of control, incapable of moving data from vendor to cloud, etc. [12].

9.7 Conclusions

This book chapter explains the various aspects of cloud computing services in the food industries. Cloud computing provides multiple service access to food industries despite several limitations for which food industries pay for the services it accesses. Services such as web-related infrastructure and services and different applications software support to food industries. It provides data virtualization of resources, data sharing, and data filtering through other systems implemented in the food industries. Cloud computing introductions ensure low hardware and software requirements, minimum maintenance cost, easy accessibility, and flexibility are significant benefits in food industries. Food industries can improve the services offered through cloud computing technologies. Although new technologies help in different ways, it also includes some challenges such as lack of resources, security, managing cloud spend, compliance, governance, managing multiple cloud services, unplanned outages, performance, etc. This chapter introduced the impact of cloud computing implementations in food and allied industries where the cost can be minimized and improved the service's qualities.

References

1. Habib M, Hauke S, Ries S, Uhlhäuser MM (2012) Review open access trust as a facilitator in cloud computing: a survey. J Cloud Comput Adv Syst Appl 1:19. Retrieved from http://www.journalofcloudcomputing.com/content/
2. Wang P, Gao RX, Fan Z (2015) Cloud computing for cloud manufacturing: benefits and limitations. J Manuf Sci Eng 137(4):044002. https://doi.org/10.1115/1.4030209
3. Intelligence Unit from The Economist (2016) Ascending cloud the adoption of cloud computing in five industries. The Econ 16. Retrieved from https://www.eiuperspectives.economist.com/sites/default/files/EIU_AscendingcloudMBP_PDF_1.pdf
4. Noor TH, Zeadally S, Alfazi A, Sheng QZ (2018) Mobile cloud computing: challenges and future research directions. J Netw Comput Appl 115(May):70–85. https://doi.org/10.1016/j.jnca.2018.04.018
5. Cloud computing in Food Sector (n.d.)
6. Operators B (n.d.) What every F & B executive needs to know the power of cloud for
7. Cloud Computing Market Size, Share_Industry Report, 2020–2027 (n.d.)
8. Cloud Infrastructure as a Service, Worldwide.pdf (n.d.)
9. Felix Richter (2020) Chart: Amazon leads $100 Billion cloud market|Statista. Retrieved from https://www.statista.com/chart/18819/worldwide-market-share-of-leading-cloud-infrastructure-service-providers/
10. Costello K, Rimol M (2019) Gartner forecasts worldwide public cloud revenue to grow 17% in 2020. Gartner Press Release. Retrieved from https://www.gartner.com/en/newsroom/press-releases/2019-11-13-gartner-forecasts-worldwide-public-cloud-revenue-to-grow-17-percent-in-2020
11. Cloud Migration Market_Growth, Trends, Forecasts (2020–2025) (n.d.)
12. Opara-Martins J, Sahandi R, Tian F (2016) Critical analysis of vendor lock-in and its impact on cloud computing migration: a business perspective. J Cloud Comput 5(1). https://doi.org/10.1186/s13677-016-0054-z
13. Pardeshi VH (2014) Cloud computing for higher education institutes: architecture, strategy and recommendations for effective adaptation. Procedia Econ Fin 11(14):589–599. https://doi.org/10.1016/S2212-5671(14)00224-X
14. Wu D, Hugenholtz P, Mavromatis K, Pukall R, Dalin E, Ivanova NN et al (2009) Cloud computing—an overview an overview. White Paper 462(7276):1–5. https://doi.org/10.1038/nature08656
15. Ferraz F, Ribeiro F, Lima W, Sampaio C (2018) A disturbing question: what is the economical impact of cloud computing? A systematic mapping. In: 2018 IEEE 11th international conference on cloud computing (CLOUD), pp 853–856. https://doi.org/10.1109/CLOUD.2018.00120

16. Bittencourt LF, Goldman A, Madeira ERM, Nelson LS, Sakellariou R (2018) Scheduling in distributed systems: a cloud computing perspective. Comput Sci Rev 30:31–54. https://doi.org/10.1016/j.cosrev.2018.08.002
17. Kobusińska A, Leung C, Hsu C, Raghavendra S, Chang V (2018) Emerging trends, issues and challenges in internet of things, big data and cloud computing. Futur Gener Comput Syst 87:416–419. https://doi.org/10.1016/j.future.2018.05.021
18. Ali KE, Mazen SA, Hassanein EE (2018) SC. Future Comput Inf J. https://doi.org/10.1016/j.fcij.2018.09.001
19. Mahmood Z (2011) Cloud computing for enterprise architectures 3–19. https://doi.org/10.1007/978-1-4471-2236-4
20. Daniluk D, Holtkamp B (2015) Logistics mall—a cloud platform for logistics, pp 13–27. https://doi.org/10.1007/978-3-319-13404-8_2
21. Phillipov M (2017) Media and food industries: the new politics of food. Media Food Ind New Polit Food 1–259. https://doi.org/10.1007/978-3-319-64101-0
22. Deloitte, FICCI (2018) Industry 4.0 in Food Industry India Food Report (January), pp 1–19
23. Chen YH (2020) Intelligent algorithms for cold chain logistics distribution optimization based on big data cloud computing analysis. J Cloud Comput 9(1). https://doi.org/10.1186/s13677-020-00174-x
24. Sandholm T, Lee D (2014) Notes on cloud computing principles. J Cloud Comput 3(1):1–10. https://doi.org/10.1186/s13677-014-0021-5
25. RightScale (2018) RightScale 2018 state of the cloud Report™, pp 46. https://doi.org/10.1016/J.EXGER.2014.04.005
26. Top 20 Cloud Computing Issues & Challenges (Latest) (n.d.)
27. Singh Gill S (2018) Failure management for reliable cloud computing: a taxonomy, model and future directions 9615(c):1–10. https://doi.org/10.1109/MCSE.2018.2873866
28. Advantages of Cloud Computing in 2019|McAfee MVISION Cloud (n.d.) Retrieved from https://www.skyhighnetworks.com/cloud-security-blog/11-advantages-of-cloud-computing-and-how-your-business-can-benefit-from-them/
29. Application of cloud computing to learning (2010) Afr J Bus Manag 4:3217–3225
30. Cegielski CG, Allison Jones-Farmer L, Wu Y, Hazen BT (2012) Adoption of cloud computing technologies in supply chains: an organizational information processing theory approach. Int J Logist Manag 23(2):184–211. https://doi.org/10.1108/09574091211265350
31. CompTIA|2018 Trends in Cloud Computing (n.d.) Retrieved from https://www.comptia.org/content/research/2018-trends-in-cloud-computing

Chapter 10
Green Technological Approaches for Food Quality and Safety in Food Supply Chain

Owais Yousuf and Anupama Singh

Abstract In a food supply chain, food quality and safety are of paramount importance as there is a drastic change in consumers' food habits, behaviour and markets. Consumer expectations for food quality and safety during the current pandemic situation are the main driving forces behind this change. Food safety has become essential, and in order to have a safe food and to get a sustainable system of food in a food supply chain, emerging green technologies are of immense importance. In a food supply chain, having good-quality food is necessary, which is safe for consumption and free of contamination. Considering these objectives, acquiring the food safety and quality is of immense prominence and responsibility of all the stakeholders in food supply chain. Alongside this journey of food products from the field to fork, there are several risks linked to food contamination which may occur in the whole supply chain. With an aim of achieving the safety and quality in the food supply chain, stakeholders, besides following the legislations and standards have also to include the effective green technological approaches in the supply chain. Evidently, several green technological approaches have emerged which can be adapted with an intention of confirming the food quality and safety in the overall food supply chain. The aim of this chapter is to provide a concise description of the green technologies and to discuss them from the view point of food quality and safety.

Keywords Food supply chain · Food quality · Food safety · Green technological approaches

O. Yousuf (✉)
Department of Bioengineering, Integral University, Lucknow 226026, India

A. Singh
Department of Food Engineering, National Institute of Food Technology Entrepreneurship and Management, Kundli (Sonepat) 131028, India

R. S Mor et al. (eds.), *Operations and Supply Chain Management in the Food Industry*, Lecture Notes in Management and Industrial Engineering,
https://doi.org/10.1007/978-981-16-5555-5_11

10.1 Introduction

In a food chain, quality together with safety of the food products is of prime importance. For this quality assurance, green technologies need to be identified from the view point of food safety. The quality and safety starts with the production and storage of food products and continues with processing, packaging, distribution before reaching to the table of the consumers [1]. In general, food supply chain, consists of various players connected together through the farm to the fork, with an objective of producing the consumer-oriented products and an augmented travel of the food products throughout the supply chain [2]. With an increased consumer perception for the food, supply chain has turned out more complex. The consumer's criteria for food selection is ever more dependent on the food characteristics of quality, safety, taste, shelf life and added health welfares [3]. Consumers are not only becoming more conscious about these factors, but are even more aware of the factors associated with food supply chain which include practices for food production and also the side effect of the food processing on environment and human health [4]. With the intention of safeguarding the food supply chain, ample safety needs are to be investigated and monitored at various levels from the producers to the consumers. This is a vital obligation not only to protect the food from triggering different health hazards but also to minimize the post-harvest losses, which is equally significant in those circumstances where food security is susceptible [2].

With due course of time, the approaches for the food quality and safety in the food supply chain have intensified. Worldwide, many countries have resorted to the food quality and safety awareness by evolving key approaches encompassing the whole supply chain [5]. Presently, food quality and safety verification in food supply chain comprise of varied approaches which include, Hazard Analysis and Critical Control Point (HACCP), Good Hygiene Practices (GHPs) and Good Manufacturing Practices (GMPs). In addition to these safety systems, food standard for quality and safety is also certified through legislations and permissible criteria for the upper safe limits of microbial contamination or the lower limits of various food components in a commodity at various levels. These methods are in use in the food supply chain but with the increased demand of safer and quality food, need is to enhance these techniques and incorporate newer methods with their relevance to the food supply chain [2].

These apprehensions led to the evolution of different approaches of green techniques, the application of which will lead to the decreased use of energy, decline in the food contamination and waste generation throughout the food supply chain [3]. These green technological approaches can be incorporated with an intent of settling the safety and quality of the food in the overall food supply chain, leading to more nutrition, new flavours, texture, taste and an extended shelf life [6]. It is obvious that these approaches will enhance the competition among the players of supply chain to be more inventive, economical and bringing in ecological reforms which may fit in with the criteria of green concepts of sustainability [7]. Furthermore, new progressions, equipment, safety systems, packaging materials and testing procedures will bring out the advances in overall quality and safety in the food supply chain. [8].

10.2 Importance of Food Supply Chain

Food is very sensitive entity in the whole market and is always linked to various hazards and risks like temperature levels, relative humidity, inadequate storage, light, poor air quality, inappropriate post-harvest technologies and other such factors which surely influence its quality and safety. The food products passes through the supply chain, comprising of several operations which influence the raw materials procurement, processing of the material till conversion into the processed product, followed by its packaging and finally, marketing. These activities implicate all the stakeholders of the chain from the farmers to the consumers with varied interests and scopes, hence making it a gripping and challenging work [9]. However, these players are the major participants of this chain, they are, therefore, responsible for conditioning and upholding the quality and safety of the food products. Every activity in the supply chain is very important, and if somehow an error occurs at a point, the entire supply chain gets affected. Usually, the whole track of the supply chain is occupied with the challenges for all its stakeholders especially those confirming to the regulations to assure food product quality and safety and at the same time generate the trust of consumers. To avoid the chain to fall apart, it is very crucial that the participants value various factors of the food supply chain seriously. Moreover, considering the fact that food being a delicate market component, and to ensure its standard features, an effective management system is compulsory. To ensure this obligation, a well-organized food supply chain is necessity in the present-day complex and critical situation of food all over the world [10].

10.3 Importance of Food Quality and Safety

Food quality and safety is an indispensable notion in food supply chain. In the present times, consumers are more alarmed about the quality and safety of their food as they are more educated and informed about food-related issues or food scandals that have happened in the recent times. It is also observed that with the change in the life style, the consumers' demands associated to the food have also changed [11]. The necessity for the ready to eat, ready to serve and minimally processed foods has increased. This is because the customers have developed a consciousness towards their health shifting for better quality and safer food. There has been a standard transference in the approach of the consumers from quantity to the standard of the food products [12]. With these changes, now more concern remains with the quality and safety of food products. In this context, governments and the other stakeholders related to the food industry are developing and enforcing fresh legislations and other such possible schemes to upgrade the product quality in an effort not only to guarantee the safety of the products, but also to raise the consumer's awareness of their efforts [13]. It is noteworthy that these regulations and criterions are not enforced at the government level only, but the other components of the food supply chain are also responsible for the same [14]. It is indeed very important for the food industry to improve their overall logistic performance in the supply chain with the latest developments which will surely make the food supply chain highly efficient capable in ensuring the total food quality and safety.

10.4 Food Quality and Safety Challenges

Food sector, in the whole world, is facing challenges in terms of quality and safety of food products. Nowadays, food supply track has grown very complex with the upsurge in the number of producers, suppliers, processes, products and much more. The longer the food supply chain, the more chances for things to go wrong which results in adverse health consequences for consumers. Proliferation of food supply chain pooled with the change climate has led to larger incidences of food-borne illness which has caused compulsion of identifying the evolving food quality and safety concerns [2]. It has also been witnessed that plenty of gaps persist in the application of already existing laws and regulations which also adds to the pile of food quality and safety concerns in the whole food supply chain. New food production and resource approaches offer the possibility of outbreak and a change in the epidemiology for large widespread food-borne disease in food safety systems [6]. Therefore, the crucial feature to protect the food at all the crossroads of the food supply chain is through the approaches ensuring an improved quality and safety of food. This integrated approach in the food supply chain should be utilized to prevent risk and encourage and implement orders so as to achieve the overall quality and safety of food products in the whole chain.

10.5 Need for Technological Approach

Technological approaches are different for nearly every food product in the complete supply chain. Official support is often required for the consumer's concerns regarding food safety and solving the issues of food quality. With rapid distribution methods in the global delivery of food, serious public health risks and food hazards in some parts of the world can be shifted to other parts of the world in a very small time [11]. Customers assume the government agencies to make sure that safe food is produced and threats to human health are minimized. But many a times, due to the lack of latest knowledge, information and resources, or other such factors, government fails in providing financial and technical support for ensuring the quality and safety of food [2]. Thus, there is an increased need for enhanced regulatory food examination and processing facilities for food control execution programmes. In general, technical interventions should be initiated in the whole supply chain for promising the security in food supply chain which in turn will safeguard the quality and safety of the food.

10.6 Green Technological Approaches in Food Supply Chain

Green technological approaches in food supply chain will lead to the pursuit of overall development and economic growth, while averting ecological deprivation, loss of biodiversity and unsanctionable use of natural resources [15]. Incorporation of the green technological approaches will address the issues pertaining to the overall food supply chain, improved food production practices, viable food processing methodologies, evolving analytical techniques, societal issues influencing consumer perceptions about techniques and significance of food supply chain in sustaining the overall attributes of food products. In the food supply chain, foods are often processed for increase in shelf life, ensuring safety, quality and nutritive properties while making them more suitable to the market [16].

In order to ensure this obligation of quality and safety in food processing sector, several primary, secondary and tertiary processing ways are being used for the processing of products and converting the whole product into highly processed food products. Furthermore, various physical, chemical and biochemical processes such as thermal treatment, minimal processing, aseptic processing, chilling, drying, fermentation and preservers have been conventionally used to boost safety and quality [11]. Generally, these processes consume a lot of energy, alternate processes consuming lesser energy and having environmentally friendly impacts must be well thought off. There is thus a mounting necessity for the new technologies to work out for these evolving issues. This is of extreme importance as these green technologies are requisite to look into the issues allied with the entire supply chain. Successful adoption and dispersal of these sustainable technologies will definitely add on their economic viability as well [17]. Ultimately, food processing at various stages is much desirable in attaining improved quality and enhanced safety of the food in the food supply chain.

10.7 Green Technologies for Assuring Food Quality and Safety

Assuring the food quality and safety of the food stuffs is an extremely important target in the food supply chain. At present, various techniques are in use for achieving these targets, but there is an immense need for the incorporation of newer and novel green techniques. Taking in to cognizance these presumptions, there are various innovative green techniques which may be used to accomplish the product quality and safety with an aim to preserve the nutrients, improve the sensory properties and augment the product's shelf life [18]. Some of the best known green techniques to assist in these intents are ozone treatment, microwave technology, ultrasound technology, pulsed electric field, ohmic heating, high-pressure processing, irradiation, pulsed light technology, and active and smart packaging [19]. With an aim of sinking the risk of contamination and ensuring an overall quality and safety of food products in the whole chain, selected best possible techniques are discussed in this section.

10.7.1 Ozone Treatment

Ozone is an active sporicidal agent and an able sanitizer with a proficient antimicrobial properties. Ozone is considered as effectual sanitizers, capable of inactivating yeasts, bacterial spores, moulds, protozoan cysts and viruses and leads to an impulsive decomposition to a non-toxic product [20]. Ozone has been found effective in almost all classes of food products during processing and storage and to bring in the safety and shelf life extension of these products. In prolonging the fruits and vegetables shelf life, ozonized water works as an efficient sanitizer. Besides inactivation of microorganisms, ozone also reduces the level of pesticides on fresh produce. Ozone treatment is very effective in decreasing the Biological oxygen demand (BOD) and chemical oxygen demand (COD) of water used in washing and processing [21]. Furthermore, ozone acts as a promising environment friendly sanitizer by rapidly decomposing to oxygen and also decreases the inorganic waste accumulation in the environment [22]. To ensure effective sanitization of the commodities, different ozone dosage is required. For example, fresh meat with high fat contents, requires more ozone than fruits and vegetables, which contain high carbohydrates and low fat content [23].

The nature of the surface of food contributes considerably to the effectiveness of ozone treatment. It is observed that the microorganisms attached intensely to the surface and those attached to areas that are not easily accessible to ozone, cannot be removed by simple rinsing in ozonized water. For that reason, when ozone is applied in food processing, it should be made sure that a good interaction between the target microorganisms and the sanitizer on the treated food takes place [24]. As per US FDA, ozone comes under the category of GRAS antimicrobial agent in dealing with raw and fresh vegetables and fruits produce. This is the simplest techniques which can be utilized so as to render food safe. Thus, application of ozone throughout the supply chain will render the products free from any contaminations which may occur during its journey through the food supply chain.

10.7.2 Microwave Technology

The Microwave technique, at present times, is deliberated as an emerging and efficient technique in the food processing which is capable of various unit operations in the food processing industry [25]. In comparison to the conventional approaches, microwave possesses the ability to save energy, hold higher rates of heating, partake very low maintenance, take less processing temperature and time, and in most of the cases is very safe and easy in handling [8]. In microwave heating, heat is produced through electromagnetic waves of certain frequencies. Most commonly operated microwave frequencies for the food products are 2450 and 915 MHz. Despite the concern of non-uniform heating which is noteworthy a technical hurdle, the process shows significant potential when combined with other techniques for improved quality and safer foods [26]. Foodstuff such as precooked foods, bread and animal feed has been treated by means of microwaves for pasteurization or sterilization. Some of the other processes which can be done using microwave are pasteurization, concentration, curing, drying, baking, freeze drying, sterilizing, tempering, cooking, thawing and drying [27]. Due to its dynamic characteristics, microwave technology can be used in various ways in the food supply chain for attaining the prospects of quality and safety of the food products.

10.7.3 Pulsed Electric Field

Pulsed electric filed (PEF) technology is an efficient food processing technique in which short electricity pulses are used for the inactivation of microbial population present in the foods while at the same time imposing minimal injurious influence on food quality [28]. To evade damage and avert off-flavour to the food commodity from heat, treatment temperature is retained as low as possible. This technology is recognized to offer enhanced food quality as it being a green technological approach which reduces the unfavourable changes in quality food and nutrition, enhances the safety and keeps the overall features of the food undamaged [29]. PEF technology's application varies widely in almost all the types of food of solid, liquid and semi-solid state. It also includes application of high-voltage pulses intended for the inactivation of microbes in dairy, poultry and other liquid foodstuffs. This technique is also applicable in baking applications, processing of juices and soups which is very effective in attaining prime quality and extended shelf life [30]. PEF like other green technologies is also accomplished for its energy and economic efficiency approaches. It is therefore an effective substitute to the conventional processing techniques in food supply chain. It is worthy to mention that PEF, being a non-thermal technique, may not be restricted to the inactivation of microbes only but can also be used in the delay of decay bringing chemical and biochemical reactions and simultaneously withholding the valuable food components while processing in the food supply chain [31].

10.7.4 High-Pressure Processing

High-pressure processing (HPP) technology is an exceptional non-thermal technology for the processing of food and for confirming overall food quality and safety, by the elimination of microorganisms [32]. HPP technology is effective in improving the taste and texture of food, enhancing the nutritional and sensorial characteristic of food besides causing an ample decline in microbial load [33]. In HPP, the food commodity is exposed to the maximum pressures of 1400 MPa, although the working pressures in the food processing ranges between 300 and 600 MPa, [34]. In the recent years, the industries using the HPP for food processing has grown up broadly as this approach is found suitable for fresh products. It is observed that the blend of mild heat and high pressure causes inactivation of more fragile bacteria [35]. Sterilization of heat subtle components like shellfish, flavourings and vitamins, while pasteurization of fruit products, vegetables, meat, pickles, sauces, yoghurt, can be done using the above technique. This technique is observed to leave no traces of toxicity, rather reduces the processing time, maintains freshness, nutrients, flavour and colour and destroys the vegetative bacteria and spores. This technique produces the uniformity of treatment throughout the food product [36]. This green technology is thus a good substitute to conventional processing methods and is very effective in shelf life extension and improve the food quality and safety in the food supply chain.

10.7.5 Pulsed Light Technology

Pulsed light technology is one of the finest commercialized replacements to the orthodox cleansing processes and own an extensive range of applications in food processing technology. The prime significance of this technology is the decontamination of food or packaging material, and alongside this, it also maintains texture, nutrients and quality of the product as well [37]. This technique provides a germicidal effect which is based on the principal of photochemical and photo thermal effect and involves high-voltage electric pulses discharge of up to 70 kV/cm into the food commodity located in in the middle of electrodes for a few seconds [38]. Pulsed light is an evolving decontamination technique and a replacement of traditional pasteurization and works effectively in decreasing the pests, pathogens and microorganisms from food for attaining the safety and quality. Besides inactivation of food surface microorganisms, this approach is also used for sterilization of packaging material and the equipment as well. Sterilization on food surface is done using high intense white light which is pulsed between 10^{-8} and 10^{-1} cycles per second [39]. It is a non-thermal treatment with higher levels of energy, treated to inactivate bacterial spores in addition to vegetative cells. It also minimizes the damaging consequence of the thermal processing and chemical action on quality attributes. Application of this method used in food processing emits one to 20 flashes per second of electromagnetic energy which is utilized in sterilization of dairy products, water, vegetables, equipment surface and cleanliness of packaging materials [40]. Pulsed light technique is projected for the future surface treatment of food products and packaging material, guaranteeing the safer food products in the whole supply chain.

10.7.6 Ohmic Heating

Ohmic heating is a green technological approach wherein the thermal energy is produced inside the food products during the processing. Contrasting to the conventional food processing technique, ohmic heating is effectual in processing food with large solid particulates. In this approach, food is located in between the equipment electrodes and an electric current is passed through it. The food product, solid or liquid, acts as an electrical resistance and gets heated up [41]. Heat is produced when electric current is conducted through the food, transforming the electrical energy into heat and destroying the microbes [42]. Application of this method in food products depends evidently on their electrical conductivity. This technique is pertinent in various unit operations like dehydration, blanching, evaporation, value addition and extraction [43]. It is presumed that with the help of ohmic heating, the expressed foods could be preserved by heating the suspended solids and liquids at diverse process streams and united later. One more property associated with this technique is that the food product swiftly attains the higher temperature after this treatment and cool off quickly as well, thus inhibiting the vitamin and nutrients loss while at the same time destruction of microorganisms is evident [44]. This green technique can be very effectively used in various processes of the food supply chain especially in baking, cooking, thawing, blanching, enhanced diffusion, electroporation, sterilization of juice, and milk [45].

10.7.7 Ultrasound Technology

Ultrasound (US) is an effective green technology applied widely in food processing operations. It is also called as supersonic technology for having the frequency greater than 20 kHz. In food processing, the ultrasound of low-frequency is engaged in the intensification of processes, while ultrasound with high frequency is utilized as an analytical technique effective for the process monitoring, quality control and safety considerations [46]. The feasibility of US technology has been extensively considered and is found to possess the abilities of enhanced competence of processes, reduced energy intake and to harvest better quality and safer food commodities [47]. Ultrasound when combined with reasonable heat aids in inactivation of microorganism. It has an immense application in drying, usually done at low temperature reducing the possibilities of degradation or oxidation in contrast to conventional methodology. Ultrasound has probable use for emulsified foods, particularly where a product's rheological assets can be enhanced by ultrasound treatment. Heat transfer between the solid heated exterior and liquid is enhanced by roughly 30–60% by engaging ultrasound technique [48]. Ultrasound is being used extensively in decontamination, emulsification, extraction, crystallization, inactivation of enzymes, viscosity variation, coating, spraying, anti-fouling and de-foaming processes [49]. This is an evolving technique which can display a praiseworthy role in the overall food quality and safety features of the food products in the food supply chain.

10.7.8 Irradiation

Persistent increase in food loss due to manifestation, contamination and spoilage, and rising distresses over food-borne diseases gave rise to emergence of the irradiation technique. This process not only assists as a conservation technique for post-harvest damages, owing to insect manifestation and premature germination, but also supports in reducing the dependency on chemical pesticides [50]. For roots and tubers, germination is the chief reason of damages. Radiation processing proposes a substitute to fumigation and certain additional treatments which is helpful in decreasing the bacteria population [51]. The irradiation effect penetrates the product and eliminates microorganisms that are found in the food products. Microbial and pathogen counts can be greatly wilted by low-dose irradiation, deprived of disturbing the sensory attributes [52]. Irradiation of vegetables and fruits is approved by US-FDA to a limit of 1 kGy for disinfection. It is considered as an efficient, well-established and a harmless processing technique [53]. Applications of this technique are very much varied and usually include decontamination of dry vegetables, spices, grains, meat, seafood, fruits, inhibition of root crop sprouts and bulbs, and sterilization of foodstuffs like functional foods, honey and mushroom. It is a favourable technique that can be utilized to boost the safety of food commodity [54]. Irradiation lengthens the shelf life of several products by the annihilation of microorganisms, and therefore, is an imperative approach in the food quality and safety of the food supply chain.

10.7.9 Active and Smart Packaging

Food packaging technology is an essential feature of the food processing sector. Food quality and safety often gets compromised due to the microbial contamination and other factors in the food supply chain. In order to avert these challenges and to impart better food safety features in the food supply chain, newer food packaging approaches play an active role. Over the time, several changes have taken place in the food packaging system which has evolved from the customary containers to the present day active and intelligent packaging systems [55]. Presently, active and smart packaging techniques are in focus of food industry which can be efficiently used for imparting quality and safety of the food products. An active packaging system not only contains the product but also acts as an active hurdle in protecting the food product from the external environmental issues. The active packaging system maintains the quality and safety of the food product, hinders adversarial effects of environment, enhances shelf life and in some cases improves the sensory characteristics of the food products [56]. Active packaging is an outstanding green technique in which the food, surroundings and packaging interrelate together to augment the quality, safety and shelf life of the food product.

In addition to active packaging, smart packaging is also an innovative technology which not only enhances the quality and the safety of food products but also keeps consumers updated about the packed food product [57]. Active and smart packaging technique is customary with the combination of various active components in the packaging material which are able to absorb or release the constituents in or out of the food product or the adjacent space. Packaging system comprises of active compounds which include absorbers of oxygen, water, ethylene, carbon dioxide and other several compounds [58]. Active packaging contains numerous gas absorbents and emitters and is considered as the significant packaging means for food products. The application of an active system having an engineered package to scavenge the molecules as O_2, CO_2 and ethylene finds use in the fruit and vegetable segment as well as the other food products throughout the supply chain [59].

10.7.10 Food Traceability

Traceability is the techniques of getting to know about the whereabouts of food product in the food supply chain. In other words traceability gives the exact documented transparency of the commodity in the food supply chain [60]. As per Codex Alimantarius Commission (CAC), "Traceability is the ability to follow the movement of the food product though the specified stages of production, processing and distribution" [61]. On the other hand, ISO standard 22005:2007 defines food traceability as a technical tool to determine the history or location of the product or its relevant components. Traceability is used to determine both forward as well as the backward linkages of food supply chain in terms of tracing and tracking [62]. Tracing determines backward process which deals with the information and record related to the origin of the product while tracking determines the forward process which deals with the identification of consumers or the end users of the product [63]. In both the cases of tracing and tracking, vital supervisory assessments are to be made in the whole supply chain for not only maintaining the quality and safety of the food products but also for the overall improvement of the supply chain [64].

Traceability is very effective in regulating and checking out the behaviour and condition of the product in the chain and will thus be very helpful in attaining the quality and safety of food products. Besides this, traceability will surely help in maintaining the transparency, reduces the risk related to food safety, food fraud identification, decreased food waste and can also manage the human errors in food supply chain [65]. In the supply chain contemporary technologies like information and communication technology, location mapping technology, sensor technology, wireless technology, and internet technology can be used by the stakeholders of the chain for accessing the food traceability. This will not only serve for achieving the quality and safety of the products but will also act as the preventive management system at times of risks and hazards which in turn will help for timely recall if needed [66]. Bar codes, sensors, QR codes and radio frequency tags are now more often used for the identification of the products, need is for the inaction of such techniques in whole chain for achieving the objectives of traceability in the food supply chain [67].

10.8 Future Outlook

In the recent years, the application of micro and nanotechnology is on an increasing trend in the food processing sector. It is therefore expected that they will play an imperative role in the whole food supply chain in the near future. Besides this, encouraging results and applications are also reported in the field of food packaging. Keeping this thing in mind, incorporation of micro and nanotechnology with the amalgamation of encapsulation technology in the food packaging sector will be of immense importance in the overall food supply chain. Another vital aspect which needs to be considered in the food supply chain is the waste management at all the points of the chain. Usually, reuse, recycle and recover are the common terms used in this context, but it is very essential to develop a suitable waste management system integrated with technological approaches for an overall sustainable development. In the supply chain, continuous valuation of the food product is very necessary. This makes traceability an important technique which is not only used for the constant check of product but will surely help in the management of the chain as well. Need is for the augmentation of traceability technique with the merger of newer systems of block chain, Internet of Things and information and communication technology. Furthermore, the regulation of these technological approaches for different food products in various supply chains will play a considerable role in standardizing the protocols for the food quality and safety in food supply chains. It is thus quiet obvious that for the overall development of an effective and efficient food supply chain, highly collaborative effort between the stakeholders of the chain in terms of product design, sourcing, logistics, storage and waste management is must. It will surely help in achieving the objective of quality and safety of the food products in the food supply chain as these approaches are often found faster, selective and sustainable then conventional ones. Application of the innovative and upcoming technological approaches in the food supply chain will not only improve the quality and safety of the food products but will surely provide better opportunities for stakeholders in achieving the better prospects in terms of economics and at the same time developing a feasible equilibrium of environment, market and the food supply chain.

10.9 Conclusion

Attainment of the feasible stability in the food supply chain along with the improved quality and safety of the food products is a tedious task. To address these apprehensions, we look towards the incorporation of the green technological approaches to swap the traditional processing techniques in the food supply chain. It is very obvious that while keeping the knowledge about the quality and safety of the food products in the food supply chain, integration of the newer green technological approaches should be appropriate with an added acquaintance of food research, regional study, traditional practices, environmental issues and the regulations linked with the supply chain. The main motive for implementing these green technological approaches is not only to guarantee the microbial decline on the commodity at various stages of supply chain but also the superior and safer standard of the food produced. Incorporating novel techniques at various stages of the food supply chain, integrated with the improved food inspection approaches, upgraded laboratory services and better food control implementation programmes will definitely lead to the guarantee of safety and quality in the complete food supply chain. It is conclusive to report that a sustainable food supply chain with the better quality and safety of the food products can be achieved with the incorporation of innovative green technological approaches in an integrated and collaborative food supply chain. With the advent of these advancements, there will be a significant improvement in the effectiveness of the food supply chain. It is also clear that the application of these green technological approaches will surely reduce the operational costs and proliferate the food supply chain flexibility. It is thus evident that the incorporation of green technological approaches combined with the standard food laws and regulations along with the emerging techniques of traceability and block chain are going to make an impact on an overall working of the food supply chain.

References

1. Kehoe B (2006) Integrating the supply chain. Mater Manag Health Care 15(8):26–29
2. Yousuf O, Titikshya S, Singh A (2018) Technological aspects for assuring food safety in food production chain 36:953–958
3. Boye JI, Arcand Y (2013) Current trends in green technologies in food production and processing. Food Eng Rev 5(1):1–17. https://doi.org/10.1007/s12393-012-9062-z
4. López-Pedrouso M, Díaz-Reinoso B, Lorenzo JM, Cravotto G, Barba FJ, Moure A, Domínguez H, Franco D (2019) Green technologies for food processing: principal considerations. In: Innovative thermal and non-thermal processing, bioaccessibility and bioavailability of nutrients and bioactive compounds. Elsevier Inc. https://doi.org/10.1016/B978-0-12-814174-8.00003-2
5. Grunert KG (2005) Food quality and safety: consumer perception and demand. Eur Rev Agric Econ 32(3):369–391. https://doi.org/10.1093/eurrag/jbi011
6. Trienekens J, Zuurbier P (2008) Quality and safety standards in the food industry, developments and challenges. Int J Prod Econ 113(1):107–122. https://doi.org/10.1016/j.ijpe.2007.02.050
7. Pingali PL (2012) Green revolution: impacts, limits, and the path ahead. Proc Natl Acad Sci USA 109(31):12302–12308. https://doi.org/10.1073/pnas.0912953109

8. Chemat F, Rombaut N, Meullemiestre A, Turk M, Perino S, Fabiano-Tixier AS, Abert-Vian M (2017) Review of green food processing techniques. Preservation, transformation, and extraction. Innov Food Sci Emerg Technol 41:357–377. https://doi.org/10.1016/j.ifset.2017.04.016
9. Turi A, Goncalves G, Mocan M (2014) Challenges and competitiveness indicators for the sustainable development of the supply chain in food industry. Procedia Soc Behav Sci 124:133–141. https://doi.org/10.1016/j.sbspro.2014.02.469
10. Bendeković J, Naletina D, Nola I (2015) Food safety and food quality in the supply chain. https://bib.irb.hr/datoteka/786708.TP_Bendekovic_Naletina_Nola.pdf
11. Yousuf O, Titikshya S, Singh A (2018) Fresh-cut fruits and vegetables: scope in developing countries and approaches to improve quality and safety 6(3):2226–2229. https://www.researchgate.net/publication/326092027
12. Suslow TV, Oria MP, Beuchat LR, Garrett EH, Parish ME, Harris LJ, Farber JN, Busta FF (2003) Production practices as risk factors in microbial food safety of fresh and fresh-cut produce. Compre Rev Food Sci Food Saf 2(1 SUPPL.):38–77. https://doi.org/10.1111/j.1541-4337.2003.tb00030.x
13. Manning L (2007) Food safety and brand equity. British Food J 109(7):496–510. https://doi.org/10.1108/00070700710761491
14. Clarke T, Boersma M (2017) The governance of global value chains: unresolved human rights, environmental and ethical dilemmas in the apple supply chain. J Bus Ethics 143(1):111–131. https://doi.org/10.1007/s10551-015-2781-3
15. Blandford D, Burrell A, Hall A, Stevens C (2011) Foreword-3 a green growth strategy for food and agriculture: preliminary report© Oecd 2011. www.oecd.org/agriculture/greengrowth
16. Boye JI, Arcand Y (2012) Erratum: green technologies in food production and processing (green technologies in food production and processing). Food Eng Ser. https://doi.org/10.1007/978-1-4614-1587-9
17. Ho JC, Shalishali MK, Tseng TL, Ang D (2009) Opportunities in green supply chain management. Coast Bus J 8(1):18–31
18. Hobbs RJ, Arico S, Aronson J, Baron JS, Bridgewater P, Cramer VA, Epstein PR, Ewel JJ, Klink CA, Lugo AE, Norton D, Ojima D, Richardson DM, Sanderson EW, Valladares F, Vilà M, Zamora R, Zobel M (2006) Novel ecosystems: theoretical and management aspects of the new ecological world order. Glob Ecol Biogeogr 15(1):1–7. https://doi.org/10.1111/j.1466-822X.2006.00212.x
19. Jiménez-Sánchez C, Lozano-Sánchez J, Segura-Carretero A, Fernández-Gutiérrez A (2017) Alternatives to conventional thermal treatments in fruit-juice processing. Part 1: techniques and applications. Crit Rev Food Sci Nutrit 57(3):501–523. https://doi.org/10.1080/10408398.2013.867828
20. Korich DG, Mead JR, Madore MS, Sinclair NA, Sterling CR (1990) Effects of ozone, chlorine dioxide, chlorine, and monochloramine on Cryptosporidium parvum oocyst viability. Appl Environ Microbiol 56(5):1423–1428. https://doi.org/10.1128/aem.56.5.1423-1428.1990
21. Sheldon BW, Brown AL (1986) Efficacy of ozone as a disinfectant for poultry carcasses and chill water. J Food Sci 51(2):305–309. https://doi.org/10.1111/j.1365-2621.1986.tb11116.x
22. Oyama ST (2000) Chemical and catalytic properties of ozone. Catal Rev Sci Eng 42(3):279–322. https://doi.org/10.1081/CR-100100263
23. Karaca H (2010) Use of ozone in the citrus industry. Ozone Sci Eng 32(2):122–129. https://doi.org/10.1080/01919510903520605
24. Naito S (1991) Studies on utilization of ozone in food preservation. Nippon Shokuhin Kogyo Gakkaishi 38(4):360–367. https://doi.org/10.3136/nskkk1962.38.360
25. Ştefănoiu G-A, Tănase EE, Miteluţ AC, Popa ME (2016) Unconventional treatments of food: microwave vs radiofrequency. Agric Agric Sci Procedia 10:503–510. https://doi.org/10.1016/j.aaspro.2016.09.024
26. Alonso AA, Banga JR (1998) Optimal control of microwave H eating using
27. Barba FJ, Roselló-Soto E, Marszałek K, Bursać Kovačević D, Režek Jambrak A, Lorenzo JM, Chemat F, Putnik P (2019) Green food processing: concepts, strategies, and tools. Green Food Process Tech 1–21. https://doi.org/10.1016/b978-0-12-815353-6.00001-x

28. Pal M (2017) Pulsed electric field processing: an emerging technology for food preservation. J Exp Food Chem 03(02):10–12. https://doi.org/10.4172/2472-0542.1000126
29. Yogesh K (2016) Pulsed electric field processing of egg products: a review. J Food Sci Technol 53(2):934–945. https://doi.org/10.1007/s13197-015-2061-3
30. Toepfl S, Heinz V, Knorr D (2006) Application of pulsed electric fields in liquid processing
31. Abbas Syed Q (2017) Pulsed electric field technology in food preservation: a review. J Nutr Health Food Eng 6(6):168–172. https://doi.org/10.15406/jnhfe.2017.06.00219
32. Amaro-Blanco G, Delgado-Adámez J, Martín MJ, Ramírez R (2018) Active packaging using an olive leaf extract and high pressure processing for the preservation of sliced dry-cured shoulders from Iberian pigs. Innov Food Sci Emerg Technol 45:1–9. https://doi.org/10.1016/j.ifset.2017.09.017
33. Balasubramaniam VM, Barbosa-Cánovas GV, Lelieveld HLM (2016) High-pressure processing equipment for the food industry. Food Eng Series 39–65. https://doi.org/10.1007/978-1-4939-3234-4_3
34. Barba FJ, Esteve MJ, Frígola A (2012) High pressure treatment effect on physicochemical and nutritional properties of fluid foods during storage: a review. Compre Rev Food Sci Food Saf 11(3):307–322. https://doi.org/10.1111/j.1541-4337.2012.00185.x
35. Lou F, Neetoo H, Chen H, Li J (2015) High hydrostatic pressure processing: a promising nonthermal technology to inactivate viruses in high-risk foods. Annu Rev Food Sci Technol 6:389–409. https://doi.org/10.1146/annurev-food-072514-104609
36. Considine KM, Kelly AL, Fitzgerald GF, Hill C, Sleator RD (2008) High-pressure processing—effects on microbial food safety and food quality. FEMS Microbiol Lett 281(1):1–9. https://doi.org/10.1111/j.1574-6968.2008.01084.x
37. Mandal R, Mohammadi X, Wiktor A, Singh A, Singh AP (2020) Applications of pulsed light decontamination technology in food processing: an overview. Appl Sci (Switzerland) 10(10). https://doi.org/10.3390/app10103606
38. Angersbach A, Heinz V, Knorr D (2000) Effects of pulsed electric fields on cell membranes in real food systems. Innov Food Sci Emerg Technol 1(2):135–149. https://doi.org/10.1016/S1466-8564(00)00010-2
39. Knorr D, Angersbach A, Eshtiaghi MN, Heinz V, Lee DU (2001) Processing concepts based on high intensity electric field pulses. Trends Food Sci Technol 12(3–4):129–135. https://doi.org/10.1016/S0924-2244(01)00069-3
40. Abida J, Rayees B, Masoodi FA (2014) Pulsed light technology: a novel method for food preservation. Int Food Res J 21(3):839–848
41. Varghese KS, Pandey MC, Radhakrishna K, Bawa AS (2014) Technology, applications and modelling of ohmic heating: a review. J Food Sci Technol 51(10):2304–2317. https://doi.org/10.1007/s13197-012-0710-3
42. Cappato LP, Ferreira MVS, Guimaraes JT, Portela JB, Costa ALR, Freitas MQ, Cunha RL, Oliveira CAF, Mercali GD, Marzack LDF, Cruz AG (2017) Ohmic heating in dairy processing: relevant aspects for safety and quality. Trends Food Sci Technol 62:104–112. https://doi.org/10.1016/j.tifs.2017.01.010
43. Kim SS, Kang DH (2015) Effect of milk fat content on the performance of ohmic heating for inactivation of *Escherichia coli* O157:H7, *Salmonella enterica Serovar Typhimurium* and *Listeria monocytogenes*. J Appl Microbiol 119(2):475–486. https://doi.org/10.1111/jam.12867
44. Roohinejad S, Koubaa M, Greiner R, Mallikarjunan K (2019) Effect of emerging processing methods on the food quality. In: Effect of emerging processing methods on the food quality. https://doi.org/10.1007/978-3-030-18191-8
45. Lee SY, Sagong HG, Ryu S, Kang DH (2012) Effect of continuous ohmic heating to inactivate *Escherichia coli* O157: H7, *Salmonella typhimurium* and *Listeria monocytogenes* in orange juice and tomato juice. J Appl Microbiol 112(4):723–731. https://doi.org/10.1111/j.1365-2672.2012.05247.x
46. Chemat F, Zill-E-Huma, Khan MK (2011) Applications of ultrasound in food technology: processing, preservation and extraction. Ultrasonics Sonochem 18(4):813–835. https://doi.org/10.1016/j.ultsonch.2010.11.023

47. Kentish S, Feng H (2014) Applications of power ultrasound in food processing. Annu Rev Food Sci Technol 5(1):263–284. https://doi.org/10.1146/annurev-food-030212-182537
48. Khan MK, Chemat F (2011) Application of low and high power ultrasound in food analysis. Anal Chem Lett 1(1):103–114. https://doi.org/10.1080/22297928.2011.10648208
49. Majid I, Nayik GA, Nanda V (2015) Ultrasonication and food technology: a review. Cogent Food Agric 1(1). https://doi.org/10.1080/23311932.2015.1071022
50. Mari M, Neri F, Bertolini P (2008) Novel approaches to prevent and control postharvest diseases of fruits. Stewart Postharvest Rev 3(6):1–7. https://doi.org/10.2212/spr.2007.6.4
51. Del Mastro NL (2011) Role of irradiation treatment in the food industry. Int J Nuclear Gov Econ Ecol 3(3):266. https://doi.org/10.1504/ijngee.2011.042213
52. Oms-Oliu G, Martín-Belloso O, Soliva-Fortuny R (2010) Pulsed light treatments for food preservation: a review. Food Bioprocess Technol 3(1):13–23. https://doi.org/10.1007/s11947-008-0147-x
53. Kume T, Furuta M, Todoriki S, Uenoyama N, Kobayashi Y (2009) Status of food irradiation in the world. Radiat Phys Chem 78(3):222–226. https://doi.org/10.1016/j.radphyschem.2008.09.009
54. Guerrero-Beltrán JA, Barbosa-Cánovas GV (2004) Review: advantages and limitations on processing foods by UV light. Food Sci Technol Int 10(3):137–147. https://doi.org/10.1177/1082013204044359
55. Müller P, Schmid M (2019) Intelligent packaging in the food sector: a brief overview. Foods 8(1). https://doi.org/10.3390/foods8010016
56. Hosseinnejad M (2014) Active packaging for food applications—a review 2(4):1174–1180
57. O'Grady MN, Kerry JP (2008) Smart packaging technologies and their application in conventional meat packaging systems. Meat Biotech 425–451. https://doi.org/10.1007/978-0-387-79382-5_19
58. Camo J, Beltrán JA, Roncalés P (2008) Extension of the display life of lamb with an antioxidant active packaging. Meat Sci 80(4):1086–1091. https://doi.org/10.1016/j.meatsci.2008.04.031
59. Ghoshal G (2018) Recent trends in active, smart, and intelligent packaging for food products. In: Food packaging and preservation. Elsevier Inc. https://doi.org/10.1016/b978-0-12-811516-9.00010-5
60. Opara LU, Mazaud F (2001) Food traceability from field to plate. Outlook Agric 30(4):239–247. https://doi.org/10.5367/000000001101293724
61. FAO (2018) Codex alimentarius, principles for traceability/product tracing as a tool within a food inspection and certification system. Cac/Gl 1(1):1–423. http://scholar.google.com/scholar?hl=en&btnG=Search&q=intitle:PRINCIPLES+FOR+TRACEABILITY+/+PRODUCT+TRACING+AS+A+TOOL+WITHIN+A+FOOD+INSPECTION+AND+CERTIFICATION+SYSTEM#0
62. Petersen A, Green D (2007) Seafood traceability: a practical guide for the U.S. industry. http://seafood.oregonstate.edu/.pdfLinks/SeafoodTraceability-APracticalGuide.pdf
63. Bechini A, Cimino MGCA, Marcelloni F, Tomasi A (2008) Patterns and technologies for enabling supply chain traceability through collaborative e-business. Inf Softw Technol 50(4):342–359. https://doi.org/10.1016/j.infsof.2007.02.017
64. Rábade LA, Alfaro JA (2006) Buyer-supplier relationship's influence on traceability implementation in the vegetable industry. J Purch Supply Manag 12(1):39–50. https://doi.org/10.1016/j.pursup.2006.02.003
65. Aung MM, Chang YS (2014) Traceability in a food supply chain: safety and quality perspectives. Food Control 39(1):172–184. https://doi.org/10.1016/j.foodcont.2013.11.007
66. Bhat R, Jõudu I (2019) Emerging issues and challenges in agri-food supply chain. Sustain Food Supply Chains Plan Des Control Interdisc Methodol 23–37. https://doi.org/10.1016/B978-0-12-813411-5.00002-8
67. Feng H, Wang X, Duan Y, Zhang J, Zhang X (2020) Applying blockchain technology to improve agri-food traceability: a review of development methods, benefits and challenges. J Clean Prod 260. https://doi.org/10.1016/j.jclepro.2020.121031

Chapter 11
Communicative Packaging Systems for Safety of Food Products

Pushpendra Kumar, Anand Kishore, Khushbu Kumari, and Rahul S Mor

Abstract The packaging system protects the food content from moisture, oxidation, biological deterioration, and physical damage and facilitates product distribution throughout the supply chain. It is of the utmost importance to select the suitable packaging material, right packaging technique, and distribution system to meet the consumer demand. The traditional packaging system facilitates the distribution system by performing the basic function of containment, protection, preservation, and communication. Nowadays, food industry and consumers are more concerned about the product's freshness and safety apart from containment and protection. This chapter entails that a communication-based packaging system is an important technique that improves or indicates product safety in the supply chain using smart devices in the packaging system. It monitors the changes in the environments and communicates the condition of food products throughout the supply chain. Smart devices may include various indicators, sensors, radio frequency identification tags, bar codes, and other techniques that improve food safety and quality. These devices establish a chain of responsibility throughout the food supply chain and build a more transparent distribution system and transmit information, rectifying the existing challenges and improving food safety.

Keywords Communicative packaging · Smart packaging indicator · Food safety · Sensors · RFID

P. Kumar · A. Kishore (✉) · R. S Mor
Department of Food Engineering, National Institute of Food Technology Entrepreneurship and Management, Kundli (Sonepat) 131028, India

K. Kumari
National Dairy Research Institute, Karnal, India

R. S Mor et al. (eds.), *Operations and Supply Chain Management in the Food Industry*,
Lecture Notes in Management and Industrial Engineering,
https://doi.org/10.1007/978-981-16-5555-5_12

11.1 Introduction

There are four primary functions of a packaging system involving containment, protection, convenience, and communication. Protection of the food materials is one of the primary functions of packaging materials. In general, all the packaging materials protect the food material from the external environment to ensure that the food product is safe and fit for consumption and is not contaminated by external materials [1]. Moreover, these functions are connected; the food packaging industry has sought additional functions to raise the regulatory body's concern and meet the consumer demands for food safety in the supply chain and storage. Food business operators implement GAP and HACCP to ensure that these hazards do not occur in any processing stages [2]. Once the product leaves the processing site, there are possibilities that these hazards will occur in food from the environment, but most of the time, these are not monitored as the facility lacks to check them. The possible safety concerns may be contamination with physical hazards, chemical changes in food, migration from the packaging materials, and the rise of microbial load. A communicative-based packaging system can play an important role in achieving a safe food supply chain. A communicative-based packaging system monitors the condition of packaged food, carries basic information during distribution, and can convey information about its quality. It has functions like sensing, recording, detecting, tracking, and communicating to facilitate decision making to extend shelf life and warn about possible problems [3]. This chapter entails various communicative-based smart packaging systems, smart packaging devices, data layers, data processing, and wireless communication network, highlighting the research involved in food product safety. These smart package devices may be attached to primary packaging or secondary packaging to communicate throughout the supply chain.

11.2 Packaging System for Microorganisms and Toxins Detector

Foodborne diseases generally occur due to contamination of food with microorganisms and are a major threat to consumers. These microorganisms produce toxins or ill effects if ingested by consumers. It requires strict monitoring of food products in the supply chain and storage. Therefore, many modifications in the packaging have taken place to identify the growth of microorganisms. The potential target molecules for microorganisms and toxins covers organic acids (lactic and acetic acids), glucose, ethanol, volatile nitrogen compounds (ammonia, dimethylamine, trimethylamine), biogenic compounds (histamine, carbon dioxide, ATP-degradation products), and sulfuric compounds, etc [4]. The decrease or increase of the target compounds is used as an indication of the quality of food. Most of the spoilage indicators are based on color changes due to microbial metabolites produced from spoilage. Packaging systems that indicate a clear visible readout can be employed in such cased for warning the consumer about safety. The pH dyes-based indicators are frequently used in the food packaging industry to detect spoilage of food products. A typical example is a colorimetric mixed-pH dye-based indicator developed by Rukchon et al. [5] to monitor skinless chicken breast spoilage on the package. The amount of CO_2 and total color difference of the indicators correlate with the spoilage of chicken. The pH dye reagents include bromothymol blue, methyl red, bromocresol green, xylenol blue, bromocresol purple, cresol red, phenol red, and alizarin which are generally used for indicators. Gas sensors may also be used for the indication of the quality of food products. The various indicators are studied for gases, but gas sensors employ changes in the sensor's electrical properties for changes in food to correlate the decay process better. A paper-based electrical gas sensor (PEGS) was developed in a study by Barandun et al. [6] for sensing water-soluble gases using cellulose paper and carbon ink. The maximum sensitivity of PEGS was for NH_3 followed by trimethylamine and CO_2, while CO and H_2S were least detectable. There was an increase in the conductance of PEGS when the concentration of NH_3 was increased. The sensor's sensitivity was affected by environmental RH. The sensor's increased conductance is attributed to an increase in the concentration of NH_3, trimethylamine, and dimethylamine, which correlated with the microbial load of the meat. The electronic nose and biosensor also can communicate the quality of food products during the supply chain.

11.3 Packaging System for Tampering

Tampering can be defined as the change in the package condition from external forces, which may be intentional or unintentional. In general, it is the question of integrity and safety of the package. Food products are carefully sealed in packaging material before dispatching them from the company. During transit, the package may get damaged from a pointed object, or a person opens it and closes it back. If a person intentionally opens a package, it would help to acknowledge that it has tampered and might not be safe. Due to these reasons, there is a need for a packaging system to detect tampering. It will ensure that the food is safe and has not been contaminated intentionally or unintentionally from the external environment. It has a wide application for MAP and other packaged food.

For advanced packaging systems, there is continuous development in this category of packaging. Gas indicators are used to detect package leaks, sealing integrity and quality of food products. Most of the gas indicators are based on color change, transparency, and pH change. The most widely used oxygen indicator is composed of redox dyes and strong reducing agents—the conventional oxygen indicator films based on synthetic and artificial components. A two-compartment arrangement was used by Won et al. [7] for developing a tampering indicator using natural compounds (laccase, guaiacol, and cysteine). Laccase in one compartment and guaiacol and cysteine were kept in wells of the indicator. The indicator containing laccase (1 U/mL), guaiacol (5 mM), and cysteine (5 mM) was evaluated at different concentrations of oxygen (1, 10, and 21), which showed a maximum color difference at 21% oxygen after 2 h of tampering. The rate of color change increased with increasing oxygen concentration. Vu and Won [8] made a colorimetric oxygen indicator film with carrageenan and redox dyes to reduce dye leakage with water. Carrageenan was employed at 0.2, 0.3, 0.4, and 0.5% in the films using dip coating. Three different redox dyes, methylene blue (MB), azure A (AA), and thionine (Th), were used for indicator films. Leaching studies of the film indicate that the dye leaching from the carrageenan films was very low compared to zein-based reference films subjected to electrostatic bonding between carrageenan (anionic) and dye (cationic). Upon UV irradiation, MB/TiO_2/glycerol/carrageenan-based films were completely bleached in 4 min with a decreased rate for films with higher concentration. The film with 0.2% carrageenan recovered in 8 h in the presence of the oxygen at ambient condition. These carrageenan-based colorimetric oxygen indicator films can be utilized in the MAP products to maintain low oxygen levels, indicating the package's poor seal or tampering. Furthermore, the kinetic indicator study is needed to understand the effect of oxygen level on the indicator recovery.

CO_2 concentration also gives an idea about the tampering or package leak. Choi and Han [9] developed a CO_2 indicator utilizing sodium caseinate and pectin solution for real-time monitoring of food quality based on transparency and pH. Sodium caseinate tends to agglomerate at lower pH, which changes its transparency. Pectin further lowers the agglomeration point as their zeta-potential and isoelectric point is changed. In the presence of CO_2, there is a reduction in the solution's pH, which can be inferred from the solution transparency. As most of the deterioration process is assisted by microbial growth in the substrate, which increases the CO_2 concentration, this solution can potentially detect the CO_2. The indicator was used to determine the quality of the kimchi at room temperature for five days. As there was the growth of *Lactic acid bacteria* during storage, the pH changed and subsequently, the transparency altered due to a rise in CO_2 concentration. In many cases, CO_2 rise also occurs due to overripe products or fermentation. A similar principle for developing CO_2 indicator was utilized by Lee and Ko [10] with whey protein isolate. Whey protein isolate solution tends to alter its turbidity due to aggregation upon a change in pH below 6.0 due to transition at isoelectric point. A lower concentration of the whey protein isolates favored a steep decrease in the pH and turbidity in a 100% CO_2 environment. A 0.3% whey protein isolate solution used in the sachet showed a pH transition from 7.05 to 5.50 and transparency from 71.1 to 12.6% in 100 min under a 100% CO_2 environment. Jung et al. [11] indicated that the pH decreased due to reaction with the CO_2 into the headspace for the solution transparency. Chitosan-based CO_2 indicator includes, which is turbid at pH 7.0 and changes to transparent at low pH due to chitosan dissolution. 2-amino 2-methyl 1-propanol (AMP) was incorporated in an indicator to control the rate of transition of chitosan dissolution. The indicator behaves significantly different at different temperatures. At 10 °C, lactic acid bacteria's growth during storage was delayed, which took 9 days to reach pH 4.0 compared to 10 °C storage. Optimum ripening of kimchi (pH 4.0–4.2 and titrable acidity 0.4–0.8%) was achieved in 1 day and 5 days at 20 and 10 °C, which directly correlated with the transparency of the indicator solution. These indicators are not exhaustively studied, but their study for different temperatures and CO_2 levels improves these indicators' functionality. Furthermore, these have a great potential to be used as an indicator to communicate about the package tempering.

11.3.1 Time–Temperature Indicator-Based Packaging System

The temperature has a significant effect on the quality and shelf life of the food product. A time–temperature indicator (TTI) is a device that is intended to record the temperature history of the product after production [12]. TTI devices are small adhesive labels integrated onto food containers or individual packages to monitor food product quality during distribution and storage. The most common TTI is based on deformation, color movement, or color development with temperature change in the supply chain. TTI generally reflects partial or complete temperature history, which can also reflect on food product safety. An elevation in the temperature during transportation accelerates the growth of microorganisms in the food product. These microorganisms pose a threat to human safety as many of them are responsible for foodborne illnesses. TTI can be classified into six categories based on their mechanism [13]:

1. Diffusion-based TTI
2. Microbial TTI
3. Enzymatic TTI
4. Polymer-based TTI
5. Photochemical TTI
6. Electronic TTI.

TTI is dependent on temperature, and its indication changes upon temperature change. Hence, it becomes an important aspect to calculate the activation energy of these TTI to implement with the type of food that matches TTI. Furthermore, there should be alignment in the endpoint of TTI with the end of food product shelf life [14].

Diffusion-based TTI incorporates material that flows depending on the temperature of the storage. During distribution and storage of perishable food products, diffusion-based TTI systems can be an important tool for monitoring microbial quality and temperature abuse. A diffusion-based TTI with isopropyl palmitate (IPP) was studied by Kim et al. [15] for the microbiological quality of non-pasteurized angelica juice during storage. The diffusion of IPP in mm was studied and modeled isothermally over a range on temperature (15, 20, 25, 30, 35 °C) and verified at 13, 23, and 33 °C. This TTI was used at non-pasteurized angelica juice at 5, 15, and 25 °C for 48 h. Microbial growth was not favored at 5 °C. 6 log cgu/mL was taken as critical and for 15 °C, 9.7 mm was covered in 36.6 h and for 25 °C, 7.2 mm was covered in 12.5 h by TTI to reach critical growth. Dynamic temperature storage study showed a distance of 7.6 mm in 27.4 h. Hence, IPP diffusion of 7 mm is set for the threshold for microbial spoilage. Isopropyl palmitate diffusion showed similar increases in microbial growth after 12 h.

Microbial-based TTI prevails over other TTI as these give more accurate results. Kim et al. [16] optimized pH-indicator to be used with lactic acid bacteria-based TTI. Out of 5 pH indicators, bromocresol green was the best based on maximum color change, ΔE and response function $F(x)$ when tested at 37 °C followed by bromocresol purple, bromophenol blue, chlorophenol red and congo red. The TTI exhibited activation energy for 4 microbial strains of lactobacillus between 103.88 and 116.35 kJ/mol indicated that it could be used for various foods depending on their spoilage characteristics.

Enzymatic TTI utilizes enzymes that react with a substrate at a temperature-dependent rate. An enzymatic TTI was developed by Brizio and Prentice [17] based on the activity of α-amylase. Starch and iodine form a dark blue color complex. This was utilized to verify that the ham was properly pasteurized or not. When starch is heated, it is converted into maltose by the activity of α-amylase and as the concentration of the starch depletes, the dark blue complex turns whitish-yellow. Hence, heating the ham for the desired time was analogous to the complex color degradation, a suitable enzymatic TTI to ensure ham's pasteurization. A concentration of 6.5% of α-amylase was optimum for use as it coincides with the ham pasteurization and even after 4 months, this TTI works fine if stored in refrigerated condition. In a recent study, Wu et al. [18] developed a TTI in which urease reacts with carbamide to produce carbamic acid and ammonia. This results in increased solution pH and phenol red, a pH-dependent dye, and its color changes from yellow to red. TTI has activation energy (Ea) of 23.05 ± 1.15 kJ/mol for a temperature range (5, 10, 15, 20, 25, and 30 °C), which can be used with foods that exhibit similar spoilage characteristics.

A polymer-based TTI employs a polymer matrix for making TTI. Lee and Shin [19] studied the polymer-based TTI to determine activation energy. The TTI was prepared from 4,4′-bis(2-benzoxazolyl) stilbene (BBS) chromophores and commercial polymer TOPAS 5013, a copolymer of ethylene and norbornene. Beyond the polymer's glass transition temperature, irreversible phase separation occurs, which aggregates the dye resulting in the change in color of TTI. 1 and 2% BBS exhibited noticeable color change above the 140 °C. Kinetic modeling of the TTI gave activation energy of 185.67 and 179.29 kJ/mol. The TTI is suitable to use for retort processing, baking, and confectionery processing purposes.

An electrical-based TTI measures electric properties such as conductivity, resistance, etc. of substances that vary with time and temperature. Wan and Knoll [20] fabricated and investigated an electrochemical TTI. This contains a conductive activation layer and hydrophilic migration layer. TTI can be activated with the help of 0.5 M NaOH filled in the temperature-sensitive film that migrates from the doping front of TTI. The length of migration fits with the conductance, which depends on the storage temperature. This can avoid the food from temperature abuses by indicating the end of shelf life upon storage.

11.3.2 Bar Codes for Packaging System

Bar codes are intended to deliver the information when scanned. Novel barcodes are introduced in the food packaging to promote the quality of food. It incorporates using chemical or enzymes which react and change its appearance concerning the reference barcode. The code change is always subject to chemical or microbiological alterations in the food. This is emerging as a promising visual indicator that can be utilized to indicate the food safety. A US patent provides innovation that a dual barcode was made to detect toxic contamination [21]. The bar code is prepared with labeled antibodies and printed on the substrate formed with antigens. As the toxins are produced upon the product storage, the toxin binds with the substrate antigens. This leads to changes in the binding with the antibodies and the barcode changes, indicating toxins or contamination that occurs upon scan. Different antigens can be employed depending on the toxin or contaminant that is to be identified. A barcode for sensing the volatile organic compounds (VOCs) in the food sample was also developed recently. A filter paper-based colorimetric sensor array was developed by Chen et al. [22], which use a mobile phone to diagnose chicken samples' freshness and safety at different temperature ranges. Barcode sensor fabricated in square shape contains Nile red (rectangle), zinc-tetraphenylporphyrin (triangle), and methyl red (circle). These three dyes change optical readout concerning emanating VOCs and pH change for the chicken meat upon storage. The picture was taken from the phone and processed in MATLAB for R, G, and B values. The intensity of the R was analyzed in the principal component analysis because of its maximum sensitivity. A significant daily and hourly change in the PCA was observed, which was distinguishable, and it can be used to predict the freshness and aging at different temperatures. This barcode sensor can be a very promising alternative for consumers as it does not require expertise to evaluate the parameters. Further development for this sensor can be employed by developing mobile software with the database in cloud storage for different temperatures for different products. The development of novel barcodes (such as toxins and metabolites-based) can further strengthen the application of intelligent food packaging. Though there is continuous development for food product safety, more novel barcodes with a wide application should be developed. Their easy interface would help consumers read out the status of safety.

11.3.3 Radio Frequency Identification-Based Packaging System

Radio frequency identification (RFID) deals with many existing food industry problems such as waste in the supply chain, empty shelves at retail outlets, temperature, theft, and recall of dangerous products subjected to foodborne illnesses [23]. Every food business operator should adopt product recall to ensure that safe food is delivering to consumers. In case of emergencies, these affected foods can be recalled and outbreaks can be stopped. The tags implemented may be active, semi-passive, or passive. These devices operate in low frequency (125–134 kHz), high frequency (13.56 MHz), ultrahigh frequency (860–960 MHz) and super high frequency (2.45–5.8 GHz), but most of the time, high and ultrahigh frequencies are employed [24]. An active RFID continuously sends the information rather than the passive. RFID employing temperature is useful in conforming to HACCP requirements [25].

RFID is meant to deliver the information from a distance irrespective of the barcodes, giving information upon scanning. In development by Nambi et al. [26], an RFID named Auburn University Detection and Food Safety (AUDFS) was made using a biosensor interface operating at 13.56 MHz. This biosensor interface was capable of identifying pathogens like *Salmonella* and *Escherichia coli* in the food. This can be used widely in the food chains to minimize foodborne illness risk due to pathogens. Hence, it may serve to deliver safe food to the consumers. RFID technology is widely used in integration with the TTI, humidity sensors, etc., directly telling about the food package's current situation. It will reduce the consumption of spoiled food and waste and better control the supply chain by monitoring and rectifying the existing challenges toward food safety and security. Huang et al. [27] conceptualized using mobile diagnosis on RFID for food product safety. RFID operating at 13.56 MHz was employed with PDA in the mobile server with remote monitoring and diagnosis server. The remote diagnosis sensor records the food product storage temperature changes using an X-chart and statistical process control. The abnormality in the temperature was communicated and classified for its severity by transmitting data between RFID and PDA attached to the phone. Various data can be fused at a time to provide the severity of the abuse. The emergence of this concept in RFID can help the various parties in the food supply chain monitor the food status with temperature abuse.

In a recent development of critical temperature indicators (CTI), integrated RFID tag microfluidics technology was used [28]. The indicator functionality depends upon a solvent melting point. Dimethyl sulfoxide with erythrosine B is employed for horizontal and vertical CTI to understand the working of solvent-based indicators. Microfluidic technology-based CTI, when integrated with passive RFID tag operating at 13.56 MHz with a reading range of 10 cm, using hexadecane as solvent (critical temperature 19 °C), sends a signal to the attached PC/tablet interface as the critical temperature is reached with the help of the reader. The RFID signal is based on the change in the system's resistance from 1 to 8 kΩ as hexadecane melts. This microfluidic CTI integrated RFID can be utilized for the supply chains with different critical temperatures using different solvents. Another temperature and humidity measurement-based passive RFID was developed by Le et al. [29] for analyzing the freshness of the packaged vegetables. The reader can read the tag from a distance of 30 cm for temperature and humidity values. This can help determine the freshness of the packaged vegetables. RFID employed in the food system can be very helpful in continuously providing information at the interface. This can be used to assess the food products' quality and safety in the food supply chain. Moreover, various losses concerning quality and microbial growth can be controlled with conveyed information.

11.4 Conclusion

The food industry, consumers, and regulatory bodies are concerned about the safe food and transparent supply chain system for food products. In this sense, a new packaging system based on smart devices will continue to evolve to improve product traceability and reduce food waste. The communicative-based packaging system has evolved to provide better information at an optimized cost in the food supply chain and storage. In this packaging system, inexpensive labels or tags of smart devices should be attached to the package to monitor changes in food packages and communicate the food product conditions throughout the supply chain. It is concluded that smart devices-based food packaging is gaining importance for consumers, industry, and researchers due to its potential to provide information to achieve the desired benefit in food quality and safety enhancement. Various indicators, sensors, tags, and bar codes are developing in recent years, which are being integrated into the packaging systems to improve the food supply chain management. This type of packaging could add several benefits to the system by enabling real-time monitoring, integrity, and traceability of the products while moving along the supply chain. Future research should direct on the manufacturing and deployment of smart devices in packaging systems to enable a wider application to improve the food supply chain.

References

1. Robertson GL (2016) Food packaging: principles and practice. CRC Press
2. Kleter GA, Marvin HJP (2009) Indicators of emerging hazards and risks to food safety. Food Chem Toxicol 47:1022–1039
3. Lee DS, Yam KL, Vanni LP (2017) Food packaging science and technology. CRC Press, London
4. Han JH (2005) innovations in food packaging. Elsevier Academic Press
5. Rukchon C, Nopwinyuwong A, Trevanich S, Jinkarn T, Suppakul P (2014) Development of a food spoilage indicator for monitoring freshness of skinless chicken breast. Talanta 130:547–554
6. Barandun G, Soprani M, Naficy S, Grell M, Kasimatis M, Chiu LC, Ponzoni A, Guder F (2019) Cellulose Fibers Enable Near-Zero-Cost Electrical Sensing of Water- Soluble Gases. ACS Sensors 4:1662−1669
7. Won K, Jang NY, Jeon J (2016) A natural component-based oxygen indicator with in-pack activation for intelligent food packaging. J Agric Food Chem 64(51):9675–9679
8. Vu CHT, Won K (2014) Leaching-resistant carrageenan-based colorimetric oxygen indicator films for intelligent food packaging. J Agric Food Chem 62(23):7263–7267
9. Choi I, Han J (2018) Development of a novel on-off type carbon dioxide indicator based on interaction between sodium caseinate and pectin. Food Hydrocolloids 80:15–23
10. Lee K, Ko S (2014) Proof-of-concept study of a whey protein isolate based carbon dioxide indicator to measure the shelf-life of packaged foods. Food Sci Biotechnol 23(1):115–120
11. Jung J, Lee K, Puligundla P, Ko S (2013) Chitosan-based carbon dioxide indicator to communicate the onset of kimchi ripening. LWT-Food Sci Technol 54:101–106
12. Giannakourou MC, Koutsoumanis K, Nychas GJE, Taoukis PS (2005) Field evaluation of the application of time temperature integrators for monitoring fish quality in the chill chain. Int J Food Microbiol 102(3):323–336. https://doi.org/10.1016/j.ijfoodmicro.2004.11.037

13. Dan W, Juzhong P, Jianchu C, Xingqian Y, Donghong L (2014) Dynamic research of time-temperature indicating system based on alkaline lipase. Appl Mech Mater 469:422–427
14. Wanihsuksombat C, Hongtrakul V, Suppakul P (2010) Development and characterization of a prototype of a lactic acid-based time-temperature indicator for monitoring food product quality. J Food Eng 100:427–434
15. Kim JU, Ghafoor K, Ahn J, Shin S, Lee SH, Shahbaz HM, Shin H-H, Kim S, Park J (2016) Kinetic modeling and characterization of a diffusion-based time-temperature indicator (TTI) for monitoring microbial quality of non-pasteurized angelica juice. LWT Food Sci Technol 67:143–150
16. Kim MJ, Jung SW, Park HR, Lee SJ (2012) Selection of an optimum pH-indicator for developing lactic acid bacteria-based time-temperature integrators (TTI). J Food Eng 113:471–478
17. Brizio APDR, Prentice C (2015) Development of aa new time temperature indicator for enzymatic validation of pasteurization of meat products. J Food Sci 80(6):M1271–M1276
18. Wu D, Wang Y, Chen J, Ye X, Wu Q, Liu D, Ding T (2013) Preliminary study on time-temperature indicator (TTI) system based on urease. Food Control 34:230–234
19. Lee BS, Shin HS (2012) Polymer-based time-temperature indicator for high temperature processed food products. Food Sci Biotechnol 21(5):1483–1487
20. Wan X, Knoll M (2016) A new type of TTI based on electrochemical pseudo transistor. J Food Eng 168:79–83
21. Goldsmith RM, Goldsmith CH, Woodaman JG (2002) Duel bar code for food contamination detection. United States Patent, US 6,479,016 B1
22. Chen Y, Fu G, Zilberman Y, Ruan W, Ameri SK, Zhang YS, Miller E, Sonkusale SR (2017) Low cost smart phone diagnostics for food using paper-based colorimetric sensor arrays. Food Control 82:227–232
23. Harrop P (2012) Radio-frequency identifiction (RFID) for food and beverage packaging applications. In: Emerging food packaging technologies. Woodhead Publishing, pp 153–174
24. Bibi F, Guillaume C, Gontard N, Sorli B (2017) A review: RFID technology having sensing aptitudes for food industry and their contribution to tracking and monitoring of food products. Trends Food Sci Technol 62:91–103
25. Kumar P, Reinitz HW, Simunovic J, Sandeep KP, Franzon PD (2009) Overview of RFID technology and its applications in the food industry. J Food Sci 74(8):R101–R106
26. Nambi S, Nyalamadugu S, Wentworth SM, Chin BA (2003) Radio frequency identification sensors. In: Proceedings of the 7th world multi-conference on systemics, cybernetics and informatics, pp 386–390
27. Huang H-P, Chen C-S, Chen T-Y (2006) Mobile diagnosis based on RFID for food safety. In: 2006 IEEE international conference on automation science and engineering, pp 357–362
28. Lorite GS, Selkälä T, Sipola T, Palenzuela J, Jubete E, Viñuales A, Cabañero G, Grande HJ, Tuominen J, Uusitalo S, Hakalahti L (2017) Novel, smart and RFID assisted critical temperature indicator for supply chain monitoring. J Food Eng 193:20–28
29. Le GT, Tran TV, Lee H-S, Chung W-Y (2016) Long-range batteryless RF sensor for monitoring the freshness of packaged vegetables. Sens Actuators, A 237:20–28

Chapter 12
Entrepreneurial Opportunities in Agri-Food Supply Chains

Anupama Panghal, Tribhuvan Nath, and Priyanka Vern

Abstract Agri-food supply chains are a dynamic system of networks and intermediaries that coordinate at required locations to make food available. The twenty-first century agri-food supply chains are advancing and becoming more efficient with the technological developments and acceleration of globalization. The increasing complexity and uncertainty associated with agri-food markets and agri-food supply chains offer entrepreneurial opportunities to potential entrepreneurs to exploit it. The various entrepreneurship models are evolving in agri-supply chains, owing to an entrepreneur's imaginative and inventive explorations to tap new opportunities. This chapter discusses the entrepreneurial opportunities in the agri-food supply chain, highlighting the entrepreneurial challenges and innovative business models. It also provides an entrepreneurial policy framework and guidelines for developing agri-food entrepreneurship. Some relevant case studies of successful entrepreneurs associated with agri-food supply chains were also included.

Keywords Agri-food supply chain · Entrepreneurial opportunities · Agri-food entrepreneurship · Policy framework

12.1 Introduction

Agri-food supply chains consist of a sequence of processes and a set of actors providing the commodity or financial or legal inputs toward making the product available to the consumer from its point of production [1]. Agri-food supply chains play a crucial role in agricultural economies in providing producers access to markets that integrate with the forward flow of products through the processes of value addition, processing, distribution, and retail sale to the ultimate consumers. An

A. Panghal · P. Vern (✉)
Department of Food Business Management and Entrepreneurship Development, National Institute of Food Technology Entrepreneurship and Management (NIFTEM), Sonipat, Haryana, India

T. Nath
Institute of Management Studies, RGSC, Banaras Hindu University, Varanasi, Uttar Pradesh, India

R. S Mor et al. (eds.), *Operations and Supply Chain Management in the Food Industry*, Lecture Notes in Management and Industrial Engineering,
https://doi.org/10.1007/978-981-16-5555-5_13

efficient food supply chain is crucial for minimizing food wastages, addressing food safety and quality, achieving food security, and the success in agri-food businesses [2]. Agri-food supply chains in many developing countries have faced several changes and challenges in recent years in terms of globalization, commodification for being cost-effective, and consolidation across all supply chain levels [3]. Agri-food entrepreneurs can come up with innovative strategies for providing solutions and strengthening agri-food supply chains. However, entrepreneurship in the agri-food sector was not strongly promoted in developing countries including India. Considering the importance of entrepreneurship as vital to job creation, rural economic growth, and ensuring the progress of thriving agri-food SMEs, the promotion of agri-food entrepreneurship is crucial. The entrepreneurship development can also help to identify practical agribusiness strategies to strengthen agri-food supply and tackle many social and economic problems of rural areas, including sustainability and climate change. However, there are many challenges for entrepreneurship development in the agri-food sector. Therefore, a strong need was felt for developing an actionable policy framework for promoting agri-food entrepreneurship along the food supply chain. This study discusses the entrepreneurial opportunities along the agri-food supply chain, highlighting the entrepreneurial challenges, and provides a summary of some innovative business models. It also develops an entrepreneurial policy framework and guidelines for developing agri-food entrepreneurship.

12.2 Challenges for Entrepreneurs in Agri-Food Supply Chains

The agri-food supply chains in many developing countries continue to be plagued with multiple problems at every stage of the chain (Fig. 12.1). At the input supply stage, the prominent problems as highlighted in many researches were poor seed replacement rate (SRR), poor adoption of improved variety of crops, low seed germination rate, and shortage of quality inputs in the market during crop season [4]. In the production stage, the limited awareness and technical knowledge among producers about modern agriculture practices (such as organic farming, poly house horticulture, and medicinal plants), poor access and adoption to farm technology, crop disease, and poor adoption of technology in dairy, fisheries, poultry causing significant losses in quality and production volume in agriculture [5, 6]. At the food processing stage, the major problems were low food processing level, lack of sorting/grading of farm produce, high post-harvest losses, poor-quality control practices, and limited/poor packaging/labeling/branding of food products [7, 8]. Lastly, at the logistics and distribution stages, the prominent problems were inaccessibility of market, limited bargaining power of poor farmers, less share of farmers in consumer rupees, poor transportation facility, limited cold storages and warehouse, inefficient distribution channels, and less food export [4, 9].

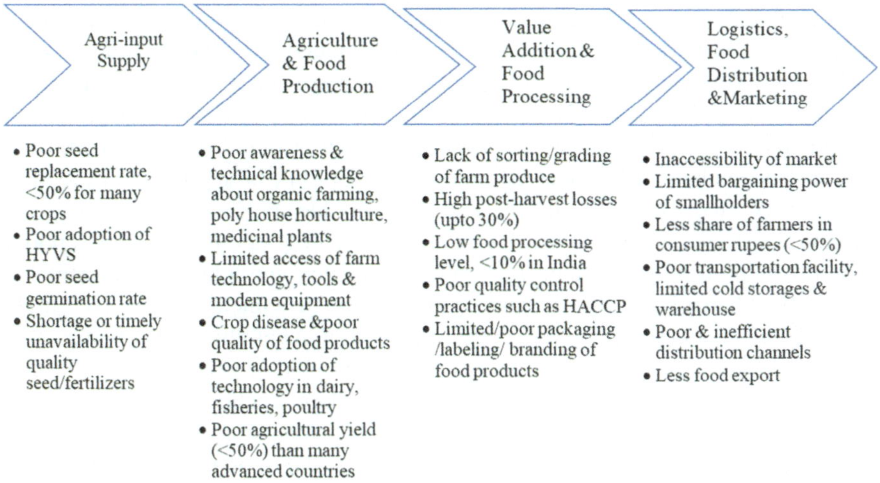

Fig. 12.1 Problems at different stages of the Agri-food supply chain. *Source* Author

The above problems are more prominent in South Asia and Sub-Saharan Africa where the awareness, availability, and adoption of promising high yielding technology (high yielding variety seed, chemical fertilizer, irrigation system) have been far from the many producers particularly smallholders due to ineffective input supply system and inadequate extension services [10, 11]. Also, the provision of technologies of high-tech production including poly house horticulture, organic farming, medicinal plants, fisheries, poultry farming is still limited to the big farmers [12, 13].

Transformation of farm produce through value addition, and food processing is also very minimal, hindering their export performance. For example, India despite being one of the primary food-producing countries with lead producers in rice, wheat, fruits, vegetable, and milk; the share in processed food in the world had just about 2% of the aggregate global exports [14]. Most of the agri-food supply chains in India are unorganized [15]. The unorganized agri-food supply chains fail to transport, store, and distribute food produces effectively. Also, limited cold chain facilities, poor warehouse conditions, traffic delays, and less organized marketing channels are common problems causing huge wastages of food products.

As various problems mentioned in Fig. 12.1, the new thinking and unique ideas can lead to innovative solutions to resolve the problems. Entrepreneurs are widely known for their ability to conceive innovative ideas to develop unique and viable solutions to satisfy customer needs that lead to growth, profitability, and business sustainability [16]. However, entrepreneurship comes with a host of challenges. A large number of startups are founded every year, but only a handful of them get converted into successful business entities. According to the study done by the IBM Institute for Business Value (IBV), 90% of startups in India fail within five years. The agri-food entrepreneurs and startups mainly struggle due to limited access to information, complicated licensing procedures, difficulty in access to finance, poor technical and managerial skills, lack of incubation facility and mentorships, including unrealistic expectations. According to a report on "Entrepreneurship in India" conducted by the National Knowledge Commission (NKC) through interview of entrepreneurs in Indian cities, about half (50%) of the entrepreneurs reported difficulties while seeking statutory clearances and licenses; more than 66% faced problems in filing taxes and 60% claimed to have encountered corruption. In addition to this, other problems reported were difficulty in accessing reliable information on registration procedures, finance, including other schemes. Some 56% of entrepreneurs reported for infrastructural issues—especially transport, power, and telecommunications—was a critical barrier. The above challenges are more prominent and involved in the agri-food sector and tough to overcome. Some significant challenges being faced by agri-food startups apart from the above are high cost of production, high post-harvest losses, seasonality of crops, low quality of products, lack of processable varieties of crops, lack of technical know-how, lack of technology, power interruption, supply chain infrastructure gaps, over-regulation of the agriculture sector, lack of skilled human resources, lack of training, limited cold storage, and marketing and export [17–20]. Addressing these challenges is a prerequisite for a favorable business environment in developing countries for startups in the agri-food sector and its supply chains.

12.3 Potential Business Opportunities along Agri-food Supply Chain

With changing times, the structure of supply chains is also changing. With the transformation of agriculture into technology and the market-oriented agribusiness sector, there are enormous entrepreneurial opportunities that have evolved along the agri-food value chain (Fig. 12.2). With increasing complexity and uncertainty in consumer demand and behavior has triggered changes in the food agri value chain which is leading to more value creation and differentiation in the food industry [21]. The entrepreneurs can look for their business interest for input supply, high-value crops, organic farming, high-tech farming, sorting and grading, dairy farming, fisheries, poultry, transport, distribution, e-food retailing, and food export are some potential enterprises.

Entrepreneurs can play a crucial role by bringing innovation in products, processes, technology, business model, and practices of agri-food supply chain management [22]. Also, the entrepreneurs can reallocate resources from the non-agriculture sector to the agriculture and food processing sector, including food distribution and logistics through knowledge, skills, and abilities. Provided with the right entrepreneurial policy framework and an ecosystem, the entrepreneurs can significantly contribute to rural prosperity, employment generation, productivity growth, and transforming rural communities by connecting the farm sector to the consumers in an efficient manner [2, 23]. The enterprises working in food supply chains need to

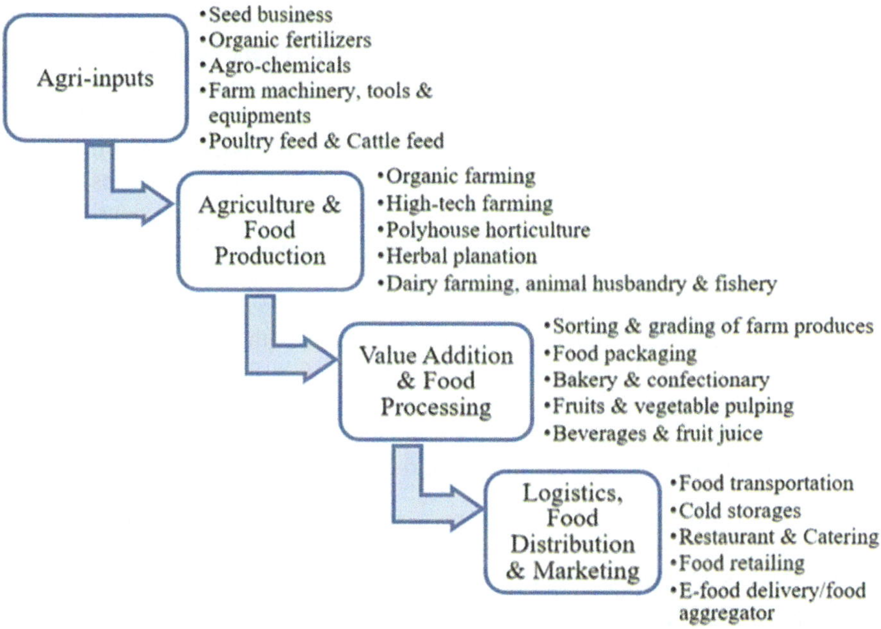

Fig. 12.2 Entrepreneurial opportunities along the agri-food supply chain. *Source* Author

face several challenges like global competition; consumer awareness and concern for food quality and safety. The enterprises need to adapt to changing business models to remain competitive [24, 25]. Organizational learning, communication, technological capabilities, and knowledge sharing must be enhanced to explore entrepreneurial opportunities in the food agribusiness [26]. The weak links in the functioning of agri-food supply chains many times offer opportunities for entrepreneurs. The supply chain participants now look for more vertical integration. Direct procurement from producers/farmers by the retailers is becoming a practice. In the past few years, various alternative models of conventional supply chains have emerged like: "direct farm selling, pick your own or u-pick; box schemes; farmers' markets; collective farmer shops, community-supported agriculture, solidarity purchasing groups, collective buying groups" [27]. All such direct marketing models integrate producers with consumers and develop close linkages between producer and consumer, which enable value sharing in the supply chains. Emergent agri-food supply chains based on the principles of lean practices and resource efficiency, with an enhanced focus on supporting the local producers [28] may be an area of interest for emerging entrepreneurs, which is profitable for them as well as beneficial for the society. Dung et al. [29] say entrepreneurial orientation benefits farmers and expands their knowledge toward agricultural entrepreneurship. Their study shows entrepreneurial orientation has a positive relationship between collaborative performance and knowledge acquisition in agri-food value chains in emerging markets. Knowledge acquisition acts as an intermediary between entrepreneurial orientation and collaborative performance. Kuckertz et al. [30] constructed an entrepreneurial opportunity space in the food value chain by mapping different sources of entrepreneurial opportunities which are change in demand, change in supply, information asymmetries, and exogenous shocks. Investors' perspective, converting and packaging, agriculture, shipping, and selling are the stages in the food value chain which provides the maximum financing. Kruja [31] discussed that collaboration between research institutions, government institutions, and the industry itself results in better entrepreneurial opportunities and act as performance enhancement in agribusiness. There are business models, emerging with new concepts like the application of blockchain technology in agri-food supply chains; artificial intelligence (A.I.)-based food business models. Entrepreneurs are also coming up with business models that support the development and promotion of local food supply chains. However, the traditional businesses of logistic providers and other 3PL services providers are also equally explored by entrepreneurs in the agri-food sector. This section will discuss a few emerging business models adopted successfully by agri-food entrepreneurs.

12.4 Policy Framework for Entrepreneurship in Agri-Food Supply Chains

Across the world, the government interest in entrepreneurship development has increased as potential solutions to boost economic growth, job creation, innovation, productivity enhancement, and the vibrant industry sector. The driving force behind this interest is the ground-level success stories and research which establish the relationship between entrepreneurship and economic upliftment [32]. The studies have revealed that government policy has a substantial influence on entrepreneurship and affects entrepreneurial activities directly or indirectly [33]. However, stimulating entrepreneurial activity requires an actionable policy framework to create an entrepreneurial ecosystem for supporting and facilitating the emergence of new entrepreneurs and their business startups. A firm entrepreneurship policy facilitates creating awareness, education, motivation, skills, finance, incubation, and mentorship of individuals willing to become entrepreneurs. The different organizations, policymakers, and researchers have suggested a different framework for entrepreneurship [34]. Food agri-supply chains are embedded with various consumer-oriented policies and producer-oriented policies. Implementation of these policies affects the behavior of various actors in an environment [35]. UNCTAD [36] has suggested a policy framework, as a collection of policies, which comprised of six areas: (i) general entrepreneurship policy, (ii) awareness and networks building, (iii) access to finance, (iv) entrepreneurship education and skills, (v) R&D and technology transfer, and (vi) regulatory environment. Lundström and Stevenson [37] have indicated six policy priorities for entrepreneurship: (i) entrepreneurship promotion, (ii) entrepreneurship education, (iii) the environment for startups, (iv) startup and seed capital financing, (v) business support measures for startups, and (vi) target group strategies.

However, an entrepreneurship policy for the agri-food sector, particularly in developing countries, needs a slightly different and a mix of above-stated policies' priorities, considering the involvement of a large number of farmers, sensitive product in terms of food quality and safety, high regulation sector and infrastructural issues. Therefore, while designing an entrepreneurial policy framework, it is essential to have a holistic view of the agri-food value chain. Authors have suggested entrepreneurship policy interventions in five areas (Table 12.1) to address all the possible aspects of agri-food entrepreneurship along the supply chain in terms of entrepreneurship promotion, awareness, motivation, education, skill development, seed financing for startups, business support, mentoring, incubation, performance measurement, and business sustainability. This policy framework is modified from Lundström and Stevenson [37] to the context of the agri-food sector.

Entrepreneurs are essential for the economy's growth. The reforms and industrial policies of the 1990s and 2000s have led to the development of the new entrepreneurship and startup landscape and the current structure. Governments, through their entrepreneurial policies, plan to overcome four policy challenges: (1) fostering the benefits of entrepreneurship as a community and employment opportunities; (2) supporting and training entrepreneurs on the skills and know-how required to thrive as an entrepreneur, along with eliminating job impediments; (3) helping young entrepreneurs; reducing obstacles and promoting access to various opportunities; (4) helping entrepreneurs through initial years of the journey by providing the necessary support, technology, elimination of information prejudice, enabling funding [38]. In 1990, industrial policy was introduced, which focused on the promotion of entrepreneurship through the campaign in rural areas. The entrepreneurship development program was conducted to train women and youths. Importance to women entrepreneurs was given. The Small Industries Development Bank of India (SIDBI) was founded on April 2, 1990, by the Indian Parliament, which currently acts as a key financial institution for the promotion, financing, and growth of the Micro, Small, and Medium Enterprises (MSME) sector. In 2000, a detailed policy package was introduced for small-scale enterprises. In 2006, the MSME Development Act came into the light which conceptualized the notion behind enterprises. Both service and manufacturing enterprises were recognized. This act has facilitated the growth of entrepreneurship in both old and young minds. The government has set in place a range of schemes to encourage and enable access to finance for aspiring entrepreneurs. To help entrepreneurs, obtain quick access to capital, the Indian government-initiated Pradhan Mantri Mudra Yojana (PMMY). This scheme was launched in 2015 to offer loans of up to 10 lakhs to non-corporate, non-farm small/micro-enterprises. In the financial year 2019–2020, around 62,237,981 PMMY loans have been sanctioned. Funds worth Rs. 337,465.13 crore have been approved for which Rs. 329,684.63 crore has been disbursed/paid to run the company (https://www.mudra.org.in/) [39].

Over the years, the Indian entrepreneurship policy framework and the startup system have emerged with the potential to improve the new venture ecosystem. Present scenario and updated regulatory framework are accommodating to entrepreneurs and open to small business ventures. But there is a holistic approach

Table 12.1 Policy framework and guidelines for entrepreneurship in agri-food supply chain

Steps in the entrepreneurial policy framework	Description: suggestive framework and guidelines
I. Entrepreneurship promotion and development of an ecosystem for entrepreneurship in the agri-food sector	• The promotion of entrepreneurship intends to create awareness among youth and other individuals about the benefits of entrepreneurship and motivate them to take up entrepreneurship as a career option. Special attention and entrepreneurial motivation should be given to farmers and rural youth toward agribusiness and food processing entrepreneurship • Entrepreneurship should be a mandatory part of the academic curriculum right from the high school level to orient the students toward entrepreneurship. The case studies of successful entrepreneurs should be included in the syllabus • The government should take strong measures for ease of doing business to create a favorable entrepreneurial climate and culture by eliminating obstacles to agri-food startups, ease the barriers to entry and exit, easy access to finance, technology transfer, IPR, incubation facility and mentorship, concession in tax in initial business years, tax reform, and infrastructure development • The different government departments/organizations (such as KVIC, DIC) involved in the promotion and development of entrepreneurship should regularly conduct entrepreneurship awareness programs in urban and rural areas. The aim is to identify potential entrepreneurs and facilitate their entrepreneurial journey. The successful entrepreneurs can be presented as a role model to motivate youth

(continued)

Table 12.1 (continued)

Steps in the entrepreneurial policy framework	Description: suggestive framework and guidelines
II. Entrepreneurial education and consortium for entrepreneurship along agri-food supply chain	• Universities, institutions, and industries specialized in agriculture, food processing, operations, and supply chain should form an entrepreneurial consortium for entrepreneurial promotion, training, and incubation of potential entrepreneurs • The higher academic institutions should mandatory open specialized centers for entrepreneurship development in their academic area, which acts as a single window for all entrepreneurial needs of potential entrepreneurs • These institutions should regularly conduct capacity building and skill development and skill upgradation for entrepreneurs. The focus should be on imparting technical and managerial skills • They should adopt some potential entrepreneurs (based on their strength) and assign academic and industry mentors and continue with them till establishment and sustainable functioning • All the government departments involved in entrepreneurship development should be linked with universities/institutions for all training and skill upgradation needs of potential entrepreneurs
III. Seed financing for agri-food startups and startup credit for agri-food entrepreneurs	• Finance is the lifeblood of business. However, for agri-food startups, seed capital is critical in the early stages as it helps get things started before the business earns any revenue. It also boosts the confidence of the entrepreneur to prove the merit of his/her idea • The government should take the step in assisting entrepreneurs in the agri-food business for seed funding to address the initial capital requirements for a startup. The online/offline platform should be created where potential entrepreneurs and investors can meet • The mentor organizations/academic institutions (who trained the entrepreneur) may be empowered for the recommendation of entrepreneurs for agri-food startup credit/micro-loan from the bank

(continued)

Table 12.1 (continued)

Steps in the entrepreneurial policy framework	Description: suggestive framework and guidelines
IV. Business support for agri-food startups—handholding, incubation, and facilitation in the establishment of agri-food enterprises	• Supporting new agri-food businesses is crucial for turning potential entrepreneurs' breakthrough ideas into successful enterprises. Importantly, startups are providing missing links in the agri-food value chain, which in turn helping the agriculture sector and farm community in realizing a higher return • However, entrepreneurs in the agri-food sector need strong support and an advisory system to not only survive but thrive. The entrepreneurial policy should extend the facilitation for access to information, technical know-how, expert advice, mentoring, and incubation support to turn their agri-food startup ideas into valuable businesses • Opening a single-window platform as a startup support center by Govt. in online mode through web-portal and the offline mode will help the entrepreneurs in establishing and sustaining their nascent agri-food enterprises
V. Measuring entrepreneurial performance and business sustainability	• Measuring the entrepreneurial performance of agro-food startup is vital to ensure the success of ventures. The performance of startups is primarily measured in terms of the degree of success in achieving the goal to generate growth and profits. The entrepreneur, including the government, should continuously watch the performance of agri-food enterprises. In case of any deviation, effort shall be made to bring it on track to avoid business failures • Sustainability is key to business success in the current business environment. Therefore, after the immediate success of agri-food startups, the business should set sustainability goals to create positive environmental and social impact apart from profits

Source Author

to creating structures like incubators, which is optimistic. Incubators support startup businesses by providing support, access to industry mentors, engaging with other founders, and most notably supplying startups with the fund to get through the survival stage. Nonetheless, a lot of changes are required at the ground level. The value of the newly built startup ecosystems does not appear to be fully percolating to entrepreneurs. Enterprises have arisen through the MUDRA scheme, and other related schemes require business support services, mentoring and constant guidance, etc. However, the ability to scale up requires qualified support, which is sometimes lacking. Some of these companies find it tough to recruit experts, and the assistance and aid they receive from government authorities are frequently inadequately trained, not well versed with recent problems, technologies, and abilities, are not per existing industry and consumer requirements. There is another scheme that has a vital role in maintaining the efficiency of the agri and food supply chain called e-NAM; the National Agricultural Market is a pan-India e-commerce portal supported by the central government and introduced by a Small Farmer Agribusiness Consortium (SFAC). These portals act as networking portals between the existing Agriculture Produce Marketing Committee (APMC), submarket yards, private markets, and other unregulated markets to bring all the agricultural markets together as one and to create an online platform. This platform recently launched two new features, *i.e.*, FPO trading module and negotiable warehouse receipt module, to e-NAM to strengthen agricultural marketing by farmers (https://www.enam.gov.in/web/) [40].

12.5 Agri-Food Supply Chain and Innovative Business Models

Supply chain innovations are crucial to any business for their survival and sustainability. This section discusses a few business models which are marking their significance in the agri-food sector (Fig. 12.3).

12.5.1 Agri-Food Traceability Models

With rising concerns about food safety and security, the traceability of food products has become an essential parameter for the success of the supply chain model of any business. Technologies like radio frequency identification (RFID) and blockchain are being used these days for product traceability [41]. RFID is used in agri-food supply chains for more than 20 years back and is an automatic identification technology based on radio frequency signals [41]. Blockchain technology is an innovation in the agri-food sector [24]. This technology is based on a distributed ledger database system, where different supply chain partners are connected through nodes on an online platform. It helps in bringing transparency in the supply chain by providing

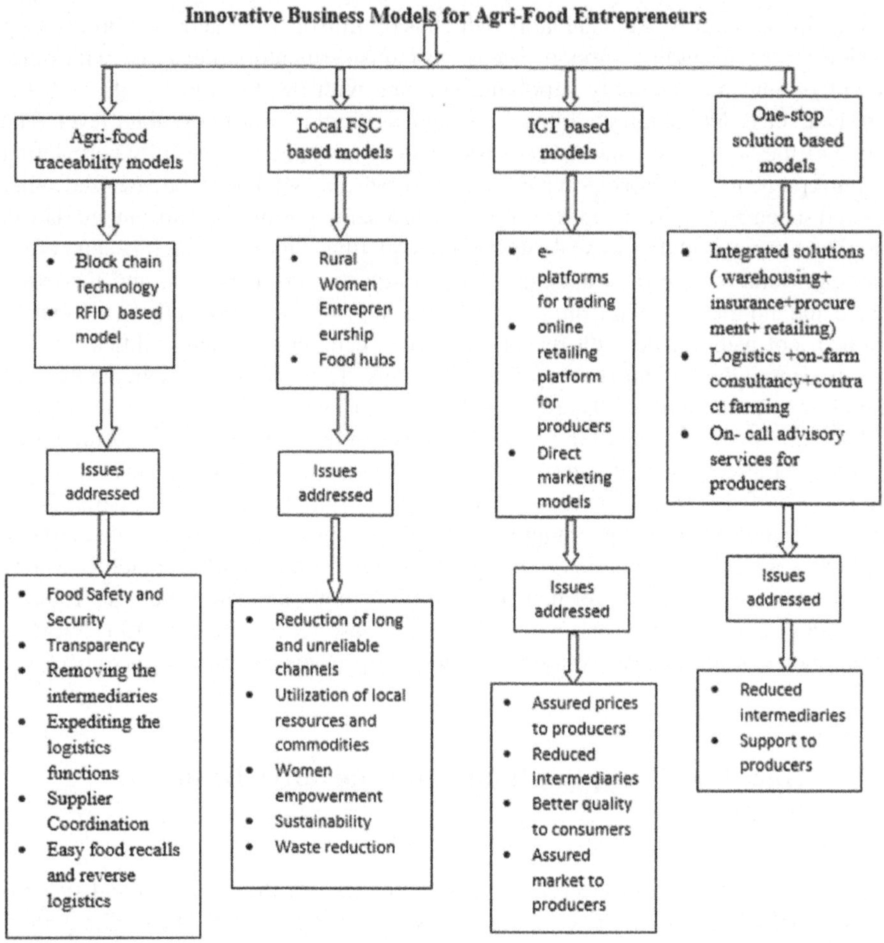

Fig. 12.3 Innovative business models for agri-food entrepreneurs. *Source* Author

complete traceability for the products [24]. The traditional agri-food supply chains have various channels and a large number of intermediaries, which complicate the logistics and distribution functions. Blockchain technology, with the help of smart contracts, helps in removing the intermediaries and expedites the logistics functions [42]. Presently, the adoption rate for blockchain technology is not much appreciated due to reasons like high costs, less knowledge, immature systems, management reluctance, etc. Still, soon with the entry of more players in this domain, blockchain technology will fetch its acceptance in agri-food systems. Galvez et al. [43] highlighted few application cases of blockchain in the agri-food supply chains like for fish (improved transaction systems); wine (complete chain management); agri-food (quality and digital identity); fruits (fairness of records); pork (reduced risk and brand protection); large food enterprises (traceability); fresh food (product information from farm to fork). Blockchain technology adoption provides 360-degree monitoring from lower management to upper management in a supply chain. It is a platform that provides easy and secure traceability, information tracking, and backtracking (Antonucci 2019).

12.5.2 Local Food Supply Chain-Based Models

The risks and the possibility of crisis at various levels in the conventional agri-food supply chains are leading to "relocalization" of supply chains in rural areas. The local food supply chains (FSC), which focus on locally grown commodities and involve fewer channels, prove to be shorter and closed-loop [27]. Along with the same lines, a term emerging and impressed upon in literature [27, 44] is "Foodhub". The "food hubs" are considered as innovative business models, where small- and medium-sized producers can collaborate for getting market access to their locally produced food. Rural women can play a significant role in promoting local food-based entrepreneurship. There are a variety of food products produced by rural women in India either at the home level or small-scale level, mostly based on traditional recipes and indigenous ingredients. Such products and business ideas have a lot of potential in urban markets. Previous studies have also highlighted the importance of local food entrepreneurship in various ways. Mars and Schau [45] have classified local food entrepreneurs into four actor types, viz. producer, organizer, purveyors, and narrators, where producer (e.g., processor and farmer) harvest and prepare product, purveyors (e.g., market vendors, restaurant owners, and retailers) and help in selling and distribution of the product. Organizers are working to enhance resources and market spaces for promotion and distribution. At last, narrators help in shaping the public image of local food within a particular region (e.g., local food journalist). Policy integration determines the success to build a framework for developing the food value chain in a rural environment [46]. The elements such as accessibility to the point of sale, sustainability in terms of economic and environmental perspective, and quality in forming a space in which coordination between actors can emerge are required in investigating alternative food networks required to collaborate to enhance each other's strengths, where local actors of territory play a strategic role in the development of food politics and consumer acceptance [47]. A strong community desired to link resources of the homeland to the consumers of the same, indifferent from their stays will always reconnect to the supply source that will also enhance consumer participation in the food value chain [48].

12.5.3 ICT-Based Models

The use of information and communication technology in agri-food supply chains is leveraged by entrepreneurs for exploiting the niche opportunities of this domain. The technologies in use range from smartphones to drones [24]. There is an emergence of electronic platforms for trading in agri-food supply chains. The concept of B2B (business to business) is also emerging in agri-food supply chains, where the entrepreneurs are providing a platform to connect producers directly to consumers and other service providers, through mobile or Internet-based platforms. Another model is by providing an online retailing platform to producers, where producers can sell their produce directly to consumers at the best possible prices. Fritz and Hausen [25] identified four electronic platform models operating in the domain of agri-food supply chains (although the efficiency is low), viz. *Neural verticals*, where transactions are performed along the complete supply chain, as per spot market environments; *downstream intermediaries*, where existing business relationships are supported at some specific later stages of the chain; *Ag output subsectors* where the focus is on agriculture-related stages of the supply chain; and *Large consortia* where adaptation is made as per the requirement of agri-food players and other participants.

12.5.4 Comprehensive One-Stop Solution-Based Models

Providing a comprehensive one-stop solution through a business model can also be an opportunity for entrepreneurs in agri-food supply chains. Some companies are providing integrated solutions under one roof like warehousing, insurance support, bulk procurement, rural retailing, etc. Similarly, companies provide logistics along with on-farm consultancy and contract farming. Producers and other supply chain partners suffer due to the unavailability of correct and updated information; entrepreneurs can fill this gap by offering producers on-call advisory services [49].

12.6 Selected Case Studies of Emerging Agri-Food Entrepreneurs

Following are the case studies of successful entrepreneurs associated with agri-food supply chains.

12.6.1 UrbanKisaan

It is an ongoing success story of a vision that started in the mind of a young entrepreneur which has now turned into the aspirations of many millennial farmers of his kind. UrbanKisaan [50] was created in 2017 by Mr. Vihari Kanukollu, Dr. Sairam Reddy, and Mr. Srinivas Chaganti. The young entrepreneur Vihari was particularly fascinated by the possibility of growing your food in a relatively little area with minimal natural resources than traditional farming practices. Mr. Vihari Kanukollu made the list of Forbes 30 under 30 Asia 2020 at the age of 26. The goal was to build a better vision for farming and feed the world in a manner that is beneficial for both the people and the earth.

UrbanKisaan's first 2500 ft^2 urban rooftop farm was built to create a hyper-local food source with a fully transparent supply chain with a minimal carbon footprint. Growing a variety of lettuces, herbs, and Indian staple leafy veggies, this farm today uses 95% less water. It produces 30 times more yields compared to traditional farms in the same area without any pesticides. The startup produces pesticide-free ingredients in a vertical hydroponic system. Hydroponics is a way to grow soil-free terrestrial plants using water-based mineral fertilizer solutions. Although farms grow plants in the soil, they use 95% less energy. UrbanKisaan manages everything from seeds to its patented, farm-controlled software framework and even empowers people to develop their own healthy, fresh, and high-quality food. This was the first time in India when someone could walk into a farm and buy the products which they can get right off the stem on their own.

UrbanKisaan in 2019 was selected to become a part of the prestigious American startup school Y-Combinator, raising its first round of seed funding at the accelerator along with other investors. There was no looking back from that eponymous moment. UrbanKisaan is currently opening its tenth retail location in Hyderabad contributing to the over 50 acres of production capacity compared to conventional farms. The smallest ideas can have the most significant impact. (*Source* Telephonic interview of the entrepreneur with the authors).

12.6.2 Ninjacart

It is India's largest fresh-produce supply chain business, a network that solves one of the world's toughest challenges through technology. They connect food producers directly to supermarkets, restaurants, and service providers using in-house applications that drive end-to-end operations. At present, their supply chain is designed to transfer 1400 tons of fresh perishables from farm to market every day in less than 12 h. The made-for-India technology and India-centric solutions have changed the way fruit and vegetables are brought to people's plates, from farms, changing the lives of farmers, companies, and customers in a significant way. Ninjacart was among the 42 most innovative startups in India in November 2018. This platform is addressing problems faced by farmers like experiencing price risk, information asymmetry about the market and consumer demands, inefficient distribution system, delay in payments. Retailers face problems such as low quality, unhygienic procurement, high market volatility. The Indian supply chain is hugely unorganized, resulting in a high incidence of food waste. Ninjacart offers solutions by removing intermediaries and gaining charge of the supply chain via automation and delivering technology-based solutions. Ninjacart provides farmers 20% more profit than the current market, one-stop selling for all goods, precise weighing and pricing, payment within 24 h, saves time. Consumers get hygienically handled products, 100% traceable to farms-improve food health, higher quality. Ninjacart is operational in seven cities presently. (*Source* Ninjacart website [51]).

12.6.3 Subzeelogic

The startup was initiated in the year 2020 by Mr. Abhishek Biswas and a team of six National Institute of Food Technology Entrepreneurship and Management (NIFTEM) students. Mr. Biswas is a business enthusiast with experience working in a variety of Asian countries such as Thailand, India, Korea, and China. He served as a trade representative and P.R. representative in the Indian Chamber of Commerce with the Thai government. He has also launched the world's first vegetarian food certification program "Sattvik Council of India".

The core idea revolves around having 99.9% non-bacterial exotic vegetables, freshly procured and packaged for sale, free of the uncertainties and difficulties of manual washing of vegetables. It all began while focusing on the challenges people face as a result of the Covid-19 pandemic and food safety and security issues. They came up with a solution to give value to both consumers and farmers by washing exotic vegetables using proprietary sanitizing technologies and making them 99.9% bacteria-free, and then packaging them in such a way as to extend product shelf life. This makes easy use of vegetables, nutritious and healthy meals, and assured consumers. Following are the steps that are included in the process of making the vegetable bacteria-free are procurement of exotic vegetables, cleaning involves the removal of dirt and soil, disinfecting using the patented technology, drying of vegetables to the optimum moisture level, packaging to increase shelf life, and testing to ensure quality and dispatch. The startup also provides guaranteed profits to linked farmers by receiving a pre-order amount of product at a fixed price. Subzeelogic [52] aims to improve the income and quality of living of farmers by providing them with improved infrastructure, support facilities, and an urban marketplace by linking Indian farmers directly to customers. There are several initiatives taken by Subzeelogic. The project began with the intention of allowing optimum use of water for washing vegetables instead of using detergents and chemicals. Many problems are faced by the Indian farmers, i.*e.,* lack of technology and lack of knowledge about marketing opportunities. At Subzeelogic, it has been a mission and an important goal to help farmers increase their income and aims to connect at least 1200 farmers in phase 1 by motivating them with Subzeelogic's value-added technology and enabling them to market their goods at higher margins with the best branding and certifications. They concentrate on generating opportunities, promising growth, having a positive impact on farmers' businesses by providing enhanced technology, supporting services, and an urban marketplace through/connecting Indian farmers directly to consumers. (*Source* Interview of the entrepreneur with the authors).

12.6.4 AgroBlock

It is a safe, clean, and sustainable agri-tech traceability platform powered by blockchain and IoT automation. It solves the issue of the traceability of the supply chain for agricultural goods that come at a premium or which are perishable since most of the supply chain companies and their suppliers face this challenge daily where the consignments are either not well protected in transit or lost. This has a considerable effect on their margins which contributes to constant follow-up, reconciliation, and additional manual inspection. The operational efficiency of the entire network will be enhanced by the implementation of a traceability scheme that gives end-to-end accountability for the supply chain and all functional activities, both in terms of saving time and reducing management and reconciliation costs. AgroBlock's ultimate goal is to play a role in encouraging the different players of agri to improve their economic partnership, by assisting agricultural companies to use the new technologies surrounding blockchain, AI, and IoT. AgroBlock [53] aims to close the gap between reliability, knowledge, and trust between the different institutions in the agricultural supply chain. These are the three cornerstones of this approach, intending to create a circular economy that will support both farmers and customers, as well as the framework intermediaries. There are various pillars of blockchain which are traceability and trust, authenticity and quality protection, real-time traceability and data reconciliation process, brand enhancement, revenue growth (increase in value), cost savings (savings in operation cost), and innovative drive. The unique selling propositions of AgroBlock are end-to-end evidential traceability, Dapp for supply chain/logistics process, supervisory input through a smart contract, IoT integration and artificial intelligence, circular approach technology, IoT automation and bio-innovation. AgroBlock is built on three pillars of the multi-party supply chain process:

Traceability It is essential to ensure that customers can always follow the trail of the supply chain in both directions, upstream and downstream, as it ensures proof that the brand procures from the sources from where it claims and the same goes for the retailers.

Visibility This is important to keep track of logistical operations as it ensures that the supply chain has not encountered any bottlenecks which are beyond the mutually accepted agreements as well as it also helps in ensuring no unsavory event has taken place during the transit.

Accountability It ensures the end product's quality. It covers to most vital aspects that the delivered product is not faked during transit and the product is viable at consumption at the end.

12.6.5 ThinkAg

It is India's first Ag-Tech platform to put together innovators, business companies, and investors to boost results in Indian food and agriculture, where fresh opportunities are increasingly opening up with increasing and evolving trends of consumption. This platform is co-founded by Mr. Ankur Capital, in collaboration with Mr. Ram Kaundinya, Mr. Raman Ahuja, and Mr. Hemendra Mathur in the year 2018. It is essential to harness these resources in the face of climate change, low productivity, supply chain inefficiencies and pollution, large-scale investment in agri-infrastructure, and technological advancement. While many entrepreneurs and innovators are using cutting-edge technologies and modern business models to address these challenges, existing players are also looking to adopt smart approaches to old businesses. Winning partnerships between stakeholders (existing entities, new entrants, innovators, and investors) to accelerate these innovations through rapid prototyping can transform the Ag-Tech landscape in India. It encourages investment in food and agriculture by enabling market validation of promising innovations. It provides AgLab, a market-linked innovation program for rapid scaling up as progressive corporates and mentors engage with budding entrepreneurs; AgInvest, a pool of industry-leading investment-ready technologies, access to experts in the industry; and AgConnect to drive innovation. ThinkAg [54] focuses on developing solutions for supply chain optimization, food waste reduction, the future generation, risk mitigation, and smart production. BigBasket, Grofers, and Ninjacart are the top three companies that raised the highest capital from the year 2014–2019 [55].

12.7 Conclusion and Policy Implications

Entrepreneurship is the primary driver of the economy across the developed and developing nations. The countries that give close attention to the promotion of entrepreneurship stand better chances of an improved economy, employment generation, innovation, industrialization, and national competitiveness. Agriculture is a dynamic and most important sector in many agricultural economies which provides a multitude of entrepreneurial opportunities along the agri-food value chain such as input supply, organic farming, high-tech farming, dairy farming, fisheries, poultry, sorting and grading, value addition and food processing, transport, distribution, e-food retailing, and food export. Most importantly, the entrepreneurs can bring innovation in products, processes, technology, and business model for economic transformation and prosperity in the agri-food sector. However, the entrepreneurs and agri-food startups face many challenges such as limited access to reliable information, complicated licensing procedure, access to finance, lack of technical skills, lack of technology, power interruption, supply chain infrastructure gaps, lack of incubation facility and mentorships apart from agriculture- and crop production-related issues. Provided with the right entrepreneurial policy framework and ecosystem, the entrepreneurs can significantly contribute to rural prosperity, employment generation, productivity growth, and transforming rural communities in developing economies. The study has suggested a robust entrepreneurial policy framework for entrepreneurship promotion, awareness, motivation, education, skill development, seed financing for startups, business support, mentoring, incubation, enterprise performance measurement, and business sustainability for creating entrepreneurial opportunities in agri-food supply chains. The study has policy implications for exploiting the entrepreneurial opportunity along the agri-food supply chain through adopting the entrepreneurial policy framework.

References

1. Cagliano R, Worley CG, Caniato FF (2016) The challenge of sustainable innovation in agri-food supply chains. Organ Sustain Effect 5:1–30
2. Knudson W, Wysocki A, Champagne J, Peterson H (2004) Entrepreneurship and innovation in the agri-food system. Am J Agr Econ 86:1330–1336
3. Mattevi M, Jones JA (2016) Traceability in the food supply chain: awareness and attitudes of UK small and medium-sized enterprises. Food Control 64:120–127
4. Ali J, Kapoor S (2009) Farmers' perception on risks in fruits and vegetables production: an empirical study of Uttar Pradesh. Agric Econ Res Rev 21:317–326
5. Raju SS, Chand R (2008) Agricultural insurance in India problems and prospects. NCAP Working Paper No. 8
6. Rao KCP (2008) Changes in dry land agriculture in the semi-arid tropics of India,1975–2004. Eur J Dev Res 20(4):562–578
7. Kumar S, Bourai VA, Kumar H (2011) Post harvest losses in pulses of Uttarakhand (a specific study of sample villages of Assan valley). Econ Aff 56(2):243–247
8. Raghavan V (2002) Preserving India's food, vol 34, no 09. McGill Reporter, McGill University
9. Rajagopal (2002) Post harvest management of mangoes with special reference to Kesar Variety in Junagadh District of Gujarat. Research Papers 1997–98, G B Pant University of Agriculture and Technology, Pantnagar, Uttaranchal
10. Kumar R, Kumar V, Jom I (2019) Study on growth and opportunity of agropreneurship in India. IUJ J Manag 7(2):117–128
11. Sandeep S (2012) Problems faced by agri entrepreneurs and remedies to solve it. J Bus Manag 3(1):23–29
12. McElwee G (2006) Farmers as entrepreneurs: developing competitive skills. J Dev Entrep 11(03):187–206
13. Pertin T (1997) Entrepreneurship as an economic force in rural development. In: Rural development through entrepreneurship, compiled and edited by Tea, REU Technical Series 41. FAO Regional Office for Europe, FAO of the United Nations, Rome
14. EXIM Bank Working Paper No. 61 (2017) International trade in processed foods: an Indian perspective
15. Samuel MV, Shah M, Sahay BS (2012) An insight into agri-food supply chains: a review. Int J Value Chain Manag 6(2):115–143

16. Suryvanci S (2012) Opportunities for agri. Entrepreneurship. ISAP
17. Awasthi D (2011) Approaches to entrepreneurship development: the Indian experience. J Glob Entrep Res 1(1):107–124
18. Bansal S, Garg D, Saini SK (2012) Agri-business practices and rural development in India—issues and challenges. Int J Appl Eng Res 7(11):1573–1577
19. Negi S (2013) Food processing entrepreneurship for rural development: drivers and challenges. In: IIM, SUSCON III third international conference on sustainability: ecology, economy and ethics. Tata McGraw Hill Education, New Delhi, pp 186–197
20. Raj SJ (2016) Problems faced by entrepreneurs: a special reference to micro and small entrepreneurs in Mumbai and Pune. Int J Bus Manag Res 6(3):57–62
21. Cucagna ME, Goldsmith PD (2018) Value adding in the agri-food value chain: research article. Int Food Agribus Manag Rev 21(3):293–316. https://doi.org/10.22434/IFAMR2017.0051
22. Berglund H (2005) Toward a theory of entrepreneurial action: exploring risk, opportunity and self in technology entrepreneurship. In: Book Doktorsavhandlingar vid Chalmers TehniskaHogskola
23. VanPraag CM, Versloot PH (2007) What is the value of entrepreneurship? A review of recent research. Small Bus Econ 29:351–382
24. Bhat R, Jõudu I (2019) Emerging issues and challenges in agri-food supply chain. In: Sustainable food supply chains. Academic Press, pp 23–37
25. Fritz M, Hausen T (2008) Electronic trade platforms in food networks: an analysis of emerging platform models and strategies. J Inf Technol Agric 3(1):26–36
26. Sachitra KMV (2019) Entrepreneurial opportunities and role of capability approach in agribusiness: evidence from Sri Lanka. Asian Res J Agric 11(1):1–11. https://doi.org/10.9734/arja/2019/v11i130051
27. Berti G, Mulligan C (2016) Competitiveness of small farms and innovative food supply chains: the role of food hubs in creating sustainable regional and local food systems. Sustainability 8(7):616
28. Anastasiadis F, Poole N (2015) Emergent supply chains in the agri-food sector: insights from a whole chain approach. Supply Chain Manag Int J
29. Dung TQ, Bonney LB, Adhikari RP, Miles MP (2020) Entrepreneurial orientation, knowledge acquisition and collaborative performance in agri-food value-chains in emerging markets. Supply Chain Manag 25(5):521–533. https://doi.org/10.1108/SCM-09-2019-0327
30. Kuckertz A, Hinderer S, Röhm P (2019) Entrepreneurship and entrepreneurial opportunities in the food value chain. Npj Sci Food 3(1):1–5. https://doi.org/10.1038/s41538-019-0039-7
31. Kruja A (2020) Entrepreneurial orientation, synergy and firm performance in the agribusiness context: an emerging market economy perspective. Central Eur Bus Rev 9(1):56–75. https://doi.org/10.18267/J.CEBR.229
32. Mandysová I (2012) Creating a context as the base for entrepreneurial support policy. Scientific Papers of the University of Pardubice, Series D, Faculty of Economics and Administration, Issue 24 (2/2012).
33. Akinyemi FO, Adejumo OO (2018) Government policies and entrepreneurship phases in emerging economies: Nigeria and South Africa. J Glob Entrep Res 8(35):1–18
34. Cassim S, Soni P, Karodia A (2014) Entrepreneurship policy in South Africa. Oman Chap Arab J Bus Manag Rev 3:29–43
35. Borsellino V, Schimmenti E, El Bilali H (2020) Agri-food markets towards sustainable patterns. Sustainability 12(6):2193. https://doi.org/10.3390/su12062193
36. UNCTAD (2009) Key components of entrepreneurship and innovation policy frameworks. Trade and Development Board. https://unctad.org/en/Docs/ciimem1d6_en.pdf
37. Lundstrom A, Stevenson L (2005) Entrepreneurial policy—theory and practice. Springer, Boston, MA
38. Shankar R, Nithyananda K (2017) Developing entrepreneurship policy in India—looking for order in chaos. In: Conference: twelfth biennial conference on entrepreneurship, vol 1. Entrepreneurship Development Institute of India Ahmedabad
39. MUDRA. https://www.mudra.org.in/. Accessed on 16.08.20

40. e-NAM. https://www.enam.gov.in/web/. Accessed on 15.08.20
41. Tian F (2016) An agri-food supply chain traceability system for China based on RFID and blockchain technology. In: 2016 13th international conference on service systems and service management (ICSSSM). IEEE, pp 1–6
42. Mezquita Y, Casado-Vara R, GonzÁlez Briones A, Prieto J, Corchado JM (2020) Blockchain-based architecture for the control of logistics activities: pharmaceutical utilities case study. Log J IGPL. jzaa039. https://doi.org/10.1093/jigpal/jzaa039
43. Galvez JF, Mejuto JC, Simal-Gandara J (2018) Future challenges on the use of blockchain for food traceability analysis. TrAC, Trends Anal Chem 107:222–232
44. Koch K, Hamm MW (2015) The role of values in food hub sourcing and distributing practices. J Hunger Environ Nutr 10:483–495
45. Mars MM, Schau HJ (2018) What is local food entrepreneurship? Variations in the commercially and socially oriented features of entrepreneurship in the Southeastern Arizona local food system. Rural Sociol 83(3):568–597
46. Graef F, Sieber S, Mutabazi K, Asch F, Biesalski HK, Bitegeko J, Bokelmann W et al (2014) Framework for participatory food security research in rural food value chains. Glob Food Sec 3(1):8–15
47. Barbera F, Dagnes J (2016) Building alternatives from the bottom-up: the case of alternative food networks. Agric Agric Sci Procedia 8:324–331
48. Kawharu M (2019) Reinterpreting the value chain in an indigenous community enterprise context. J Enterprising Commun 13(3):242–262
49. Verma I (2013) Agri-logistic in india: Challenges and Emerging Solutions. Yourstory. https://yourstory.com/2013/05/agri-logistics-in-india-challenges-and-emerging-solutions?utm_pageloadtype=scroll. Accessed on 26.8.2020
50. UrbanKisaan. https://www.urbankisaan.com/. Accessed on 20.08.20
51. Ninjacart. https://ninjacart.in/. Accessed on 26.8.20
52. Subzeelogic. http://subzeelogic.com/. Accessed on 29.8.20
53. AgroBlock. https://agroblock.io/. Accessed on 06.09.20
54. ThinkAg. https://www.thinkag.in/. Accessed on 01.08.20
55. ThinkAg (2020) Ag-Tech in India, Investment Landscape Report. Available at: https://a0e20efd-4fe9-4537-80f50bc1b751db89.usrfiles.com/ugd/a0e20e_da949f63da464c90a0ab4d25c4b280d5.pdf

Index

R. S Mor et al. (eds.), *Operations and Supply Chain Management in the Food Industry*, Lecture Notes in Management and Industrial Engineering,
https://doi.org/10.1007/978-981-16-5555-5

The manufacturer's authorised representative in the EU is Springer Nature Customer Service Centre GmbH, Europaplatz 3, 69115 Heidelberg, Germany. If you have any concerns regarding our products, please contact ProductSafety@springernature.com

Printed and bound by CPI Group (UK) Ltd, Croydon, CR0 4YY
30/09/2026
02225324-0004